CHEMISTRY
in the Community

Activities Workbook

CHEMCOM

A Project of the American Chemical Society

W.H. FREEMAN AND COMPANY
NEW YORK

Activities Workbook

American Chemical Society

Activities Workbook Editor: Mike Clemente
Chief Editor: Henry Heikkinen
Revision Team: Laurie Langdon, Robert Milne, Angela Powers
Revision Assistants: Cassie McClure, Seth Willis
Teacher Edition: Joseph Zisk, Lear Willis
Ancillary Materials: Regis Goode, Patricia Smith, Ruth Leonard
Fourth Edition Editorial Advisory Board: Conrad L. Stanitski (Chair), Boris Berenfeld, Jack Collette, Robert Dayton, Ruth Leonard, Nina I. McClelland, George Miller, Adele Mouakad, Carlo Parravano, Kirk Soulé, Maria Walsh, Sylvia A. Ware (*ex officio*), Henry Heikkinen (*ex officio*)
ACS: Sylvia Ware, Janet Boese, Michael Tinnesand, Guy Belleman, Patti Galvan, Helen Herlocker

W. H. Freeman

Publisher: Michelle Russel Julet
Text Designer: Proof Positive/Farrowlyne Associates, Inc.
Cover Designer: Diana Blume
Illustrations: Proof Positive/Farrowlyne Associates, Inc.
Production Coordinator: Susan Wein
Production Services: Proof Positive/Farrowlyne Associates, Inc.
Supplements and Multimedia Editor: Charlie Van Wagner
Composition: Black Dot Group
Manufacturing: R. R. Donnelley & Sons

Activities Workbook: ChemCom.
"A project of the American Chemical Society."

ISBN 0-7167-3918-6

Printed in the United States of America

Second printing, 2002

Contents

Unit 5

Unit 6

Unit 7

Preface

TO THE STUDENT

The material in the *Chemistry in the Community (ChemCom)* **Activities Workbook** is designed to help you learn chemistry. The book includes information and exercises that support your textbook.

ChemCom is structured around community issues related to chemistry rather than around specific chemical concepts. As a result, chemical concepts are presented on a "need-to-know basis." To fully understand the community issues, you must have a thorough and rich understanding of the chemical concepts involved, and you must know how to apply your learning.

By providing worksheets, problem sets, and laboratory handouts, the materials in the **Activities Workbook** can help develop and extend your understanding of the science concepts you are studying in your textbook. The **Activities Workbook** also provides exercises that will reinforce what you learn and help organize your work on labs and assessments.

TO THE TEACHER

The **Activities Workbook** consists of optional handouts and worksheets that you may choose to use to help students organize their work. You can use the enclosed activities in several ways. If this book is issued to a student, it is recommended that you remove the answer sheets from the back of the book before handing out the book.

The lab handouts reproduce the introduction and procedure sections of each *ChemCom* lab. With the handouts in hand, students will not risk damaging their textbooks in a "wet" lab situation. Lab handouts also provide student data tables, designed to help organize the results of laboratory work. The worksheets provided in the **Activities Workbook** reinforce concepts covered in the textbook. If you wish, before students begin to work on a laboratory activity, make copies of the Pre-Laboratory and Laboratory Report forms, which you will find in the *Black Line Copy Masters* teacher supplement.

Finally, at the end of the activities for each unit there are worksheets to help students with the presentation and assessment of the unit's Putting-It-All-Together activity. We hope these materials assist you as you and your students explore the fourth edition of *ChemCom*.

STUDENT NAME _____ Date _____

CHEMCOM LABORATORY SAFETY AGREEMENT FORM

Rules of Laboratory Conduct

1. Perform laboratory work only when your teacher is present. Unauthorized or unsupervised laboratory experimenting is not allowed.

2. Your concern for safety should begin even before the first activity. Before starting any laboratory activity, always read and think about the details of the laboratory assignment.

3. Know the location and use of all safety equipment in your laboratory. These should include the safety shower, eye wash, first-aid kit, fire extinguisher, fire blanket, exits, emergency warning system, and evacuation routes.

4. Wear a laboratory coat or apron and impact/splash-proof goggles for all laboratory work. Wear closed shoes (rather than sandals or open-toed shoes), preferably constructed of leather or similar water-impervious material, and tie back loose hair. Shorts or short skirts must not be worn.

5. Clear your benchtop of all unnecessary material such as books and clothing before starting your work.

6. Check chemical labels twice to make sure you have the correct substance and the correct concentration of a solution. Some chemical formulas and names may differ by only a letter or a number.

7. You may be asked to transfer some laboratory chemicals from a common bottle or jar to your own container. Do not return any excess material to its original container unless authorized by your teacher, as you may contaminate the common bottle.

8. Avoid unnecessary movement and talk in the laboratory.

9. Never taste laboratory materials. Do not bring gum, food, or drinks into the laboratory. Do not put fingers, pens, or pencils in your mouth while in the laboratory.

10. If you are instructed to smell something, do so by fanning some of the vapor toward your nose. Do not place your nose near the opening of the container. Your teacher will show the correct technique.

11. Never look directly down into a test tube; do view the contents from the side. Never point the open end of a test tube toward yourself or your neighbor. Never heat a test tube directly in a Bunsen burner flame.

12. Any laboratory accident, however small, should be reported immediately to your teacher.

13. In case of a chemical spill on your skin or clothing, rinse the affected area with plenty of water. If the eyes are affected, rinsing with water must begin immediately and continue for at least 10 to 15 minutes. Professional assistance must be obtained.

14. Minor skin burns should be placed under cold, running water.

15. When discarding or disposing of used materials, carefully follow the instructions provided.

16. Return equipment, chemicals, aprons, and protective goggles to their designated locations.

17. Before leaving the laboratory, make sure that gas lines and water faucets are shut off.

18. Wash your hands before leaving the laboratory.

19. If you are unclear or confused about the proper safety procedures, ask your teacher for clarification. If in doubt, ask!

Students exhibiting misconduct or disregard for safety during a laboratory period will be asked to leave the lab or will be subject to other disciplinary action.

By signing below, the student and parent or guardian indicate that they have read and have agreed to follow these "Rules of Laboratory Conduct." The student is expected to follow these rules as well as any additional printed or verbal safety instructions given by the teacher. This slip is to be returned by _____. If a parent or guardian has any questions, please feel free to telephone _____ at _____.

STUDENT SIGNATURE _____ DATE _____

PARENT/GUARDIAN SIGNATURE _____ DATE _____

Unit 1

A.2 LABORATORY ACTIVITY: FOUL WATER—PROCEDURE

Procedure

1. Obtain approximately 100 mL (milliliters) of foul water from your teacher. Measure its volume precisely with a graduated cylinder; record the actual volume of the water sample (with units) in your data table.

2. Examine the properties of your sample: color, clarity, odor, and presence of oily or solid regions. Record your observations in the "Before treatment" section of your data table.

Oil-Water Separation

As you know, if oil and water are mixed and left undisturbed, the oil and water do not noticeably dissolve in each other. Instead, two layers form. Which layer floats on top of the other? Make careful observations in the following procedure to check your answer.

1. Place a funnel in a clay triangle supported by a ring clamp and ring stand. Attach a rubber hose to the funnel tip.

2. Close the rubber tube by tightly pinching it with your fingers (or by using a metal pinch clamp). Gently swirl the foul-water sample for several seconds. Then immediately pour about half of the sample into the funnel. Let it stand for a few seconds until the liquid layers separate. (Gentle tapping may encourage oil droplets to break free.)

3. Carefully open the tube, releasing the lower liquid layer into an empty 150-mL beaker. When the lower layer has just drained out, quickly close the rubber tube.

4. Drain the remaining layer into another 150-mL beaker.

5. Repeat Steps 2–4 using the other half of your sample, adding each liquid to the correct beaker. Which beaker contains the oily layer? How do you know?

6. Dispose of the oily (non-watery) layer as instructed by your teacher. Observe the properties of the remaining layer and measure its volume. Record your results. Save this water sample for the next procedure.

7. Wash the funnel with soap and water.

Sand Filtration

A **sand filter** traps and removes solid impurities—at least those particles too large to fit between sand grains—from a liquid.

1. Using a straightened paper clip, poke small holes in the bottom of a paper cup.

2. Add pre-moistened gravel and sand layers to the cup. (The bottom layer of gravel prevents the sand from washing through the holes. The top layer of gravel keeps the sand from churning up when the water sample is poured in.)

3. *Gently* pour the liquid sample to be filtered into the cup. Catch the filtrate (filtered water) in a beaker as it drains through.

4. Dispose of the used sand and gravel according to your teacher's instructions. Do *not* pour any sand or gravel into the sink!

5. Observe the properties of the filtered water sample and measure its volume. Record your results. Save the filtered water sample for the next procedure.

Charcoal Adsorption/Filtration

Charcoal **adsorbs** (attracts and holds on its surface) many substances that could give water a bad taste, odor, or cloudy appearance. A fish aquarium often includes a charcoal filter for the same purpose.

1. Fold a piece of filter paper as shown in the lab.

2. Place the folded filter paper in a funnel. Hold the filter paper in position and wet it slightly so it clings to the inside of the funnel cone.

3. Place the funnel in a clay triangle supported by a ring clamp. Lower the ring clamp so the funnel stem extends 2–3 cm (centimeters) inside a 150-mL beaker.

4. Place one teaspoon of charcoal in a 125- or 250-mL Erlenmeyer flask.

5. Pour the water sample into the flask. Shake vigorously for a few seconds. Then gently pour the liquid through the filter paper. Keep the liquid level below the top of the filter paper. Liquid should not flow through the space between the filter paper and the funnel. (Can you explain why?)

6. If the filtrate is darkened by small charcoal particles, refilter the liquid through a clean piece of moistened filter paper.

7. When you are satisfied with the appearance and odor of your water sample, pour it into a graduated cylinder. Record the final volume and properties of the purified sample.

8. Follow your teacher's suggestions about saving your purified sample. Place used charcoal in the container provided by your teacher.

9. Wash your hands thoroughly before leaving the laboratory.

A.2 LABORATORY ACTIVITY: FOUL WATER

Purpose

The purpose of this activity is to purify a sample of foul water.

DATA TABLE

	Volume (mL)	Color	Clarity	Odor	Presence of Oil	Presence of Solids
Before treatment						
After oil-water separation						
After sand filtration						
After charcoal adsorption and filtration						

Calculations

1. _____ %

2. _____ mL

3. _____ %

Data Analysis

Post-Lab Activities

1a. _____

_____

_____

b. _____

_____

_____

c. _____

_____

_____

Other uses and frequency of each				L
				L
				L
				L
Total Water Used in Three Days				L

2. _____ L Average/person/day

3. _____ Class range or average daily water use

4. _____ Class mean

   _____ Class median

   _____

5. _____

   _____

6. _____

   _____

Unit 1

B.6 SUPPLEMENT: IONS AND IONIC COMPOUNDS

Complete the data table identifying the composition of each ionic compound.

	Cation	Anion	Formula	Name
1.				Calcium oxide
2.			NaCl	
3.	NH_4^+	NO_3^-		
4.			$Cu(OH)_2$	
5.				Iron(III) sulfate
6.	K^+	SO_3^{2-}		
7.			Na_3PO_4	
8.	Pb^{2+}	Br^-		
9.				Lead(II) carbonate
10.	Al^{3+}	PO_4^{3-}		
11.				Magnesium hydrogen carbonate
12.			K_2S	
13.	Ba^{2+}	SO_4^{2-}		
14.				Zinc phosphate
15.			$FeCl_3$	

Unit 1

B.7 LABORATORY ACTIVITY: WATER TESTING—PROCEDURE

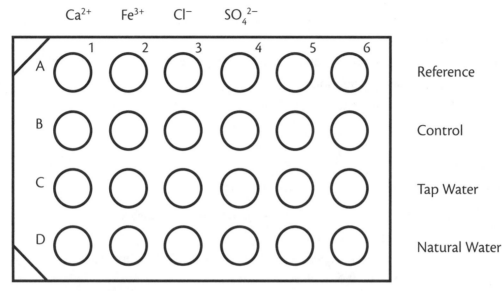

Ca^{2+} Fe^{3+} Cl^- SO_4^{2-}

A — Reference

B — Control

C — Tap Water

D — Natural Water

Procedure

The test procedures for each ion are given below. If the ion of interest is present, a chemical reaction will occur, producing either a colored solution or a precipitate. The chemical equations involved are given for each ion.

Calcium Ion Test (Ca^{2+})

$$Ca^{2+}(aq) \ + \ C_2O_4^{2-}(aq) \longrightarrow \ CaC_2O_4(s)$$
calcium ion   oxalate ion   calcium oxalate

Follow these steps for each solution (Ca^{2+} reference, control, tap water, and natural water).

1. Place 20 drops of the solution into a well of a 24-well wellplate.

2. Add three drops of sodium carbonate ($NaCO_3$) to the well.

3. Record your observations including the color and whether or not a precipitate formed.

4. Determine whether the ion is present and record your results.

5. Repeat for the remaining solutions.

6. Discard the contents of the wellplate as directed by your teacher.

Iron(III) Ion Test (Fe^{3+})

$$Fe^{3+}(aq) \quad + \quad SCN^-(aq) \quad \longrightarrow \quad [FeSCN]^{2+}(aq)$$

iron(III) ion   thiocyanate ion   iron(III) thiocyanate

Follow these steps for each solution (Fe^{3+} reference, control, tap water, and natural water).

1. Place 20 drops of the solution into a well of a 24-well wellplate.
2. Add one or two drops of potassium thiocyanate (KSCN) test solution to the well.
3. Record your observations including the color and whether or not a precipitate formed.
4. Determine whether the ion is present and record your results.
5. Repeat for the remaining solutions.
6. Discard the contents of the wellplate as directed by your teacher.

Chloride Ion Test (Cl^-)

$$Cl^-(aq) \quad + \quad Ag^+(aq) \quad \longrightarrow \quad AgCl(s)$$

chloride ion   silver ion   silver chloride

Follow the same procedure as for the Fe^{3+} ion, with the following changes:

◆ Use the Cl^- ion reference solution.
◆ In Step 2, add three drops of silver nitrate ($AgNO_3$) test solution, instead of potassium thiocyanate (KSCN) test solution.

Sulfate Ion Test (SO_4^{2-})

$$SO_4^{2-}(aq) \quad + \quad Ba^{2+}(aq) \quad \longrightarrow \quad BaSO_4(s)$$

sulfate ion   barium ion   barium sulfate

Follow the same procedures as for the Fe^{3+} and Cl^- ions with the following changes:

◆ Use the SO_4^{2-} reference solution.
◆ In Step 2, add three drops of barium chloride ($BaCl_2$) test solution, instead of potassium thiocyanate (KSCN).

Discard the wellplate as directed by your teacher.

Unit 1

B.7 LABORATORY ACTIVITY: WATER TESTING

Purpose

The purpose of this activity is to check for the presence of certain ions in water solutions.

DATA TABLE: Calcium (Ca^{2+})

Solution	Observations (color, precipitate, etc.)	Result (Is ion present?)
Reference		
Control		
Tap water		
Natural water from (source) _____		

DATA TABLE: Iron(III) (Fe^{3+})

Solution	Observations (color, precipitate, etc.)	Result (Is ion present?)
Reference		
Control		
Tap water		
Natural water from (source) _____		

DATA TABLE: Chloride (Cl^-)

Solution	Observations (color, precipitate, etc.)	Result (Is ion present?)
Reference		
Control		
Tap water		
Natural water from (source) _____		

DATA TABLE: Sulfate (SO_4^{2-})

Solution	Observations (color, precipitate, etc.)	Result (Is ion present?)
Reference		
Control		
Tap water		
Natural water from (source) _____		

Questions

1a. _____

_____

_____

b. _____

_____

_____

2. _____

_____

_____

3. _____

_____

_____

4. _____

_____

_____

Unit 1

C.1 SUPPLEMENT: SOLUBILITY AND SOLUBILITY CURVES

Using the solubility curves found in Section C of Unit 1 (pp. 46 and 62), solve the following problems.

1. What is the solubility of potassium nitrate (KNO_3) in 100.0 g of water at 80.0 °C?

   _____

2. What is the solubility of potassium chloride (KCl) in 100.0 g of water at 50.0 °C?

   _____

3. What is the solubility of sodium chloride (NaCl) in 100.0 g of water at 90.0 °C?

   _____

4. What is the minimum temperature needed to dissolve 180.0 g of KNO_3 in 100.0 g of water?

   _____

5. What is the minimum temperature needed to dissolve 35.0 g of KCl in 100.0 g of water?

   _____

6. At what temperature do KCl and KNO_3 have the same solubility?

   _____

7. How much more KCl will dissolve at 90.0 °C than at 20.0 °C?

   _____

8. If 50.0 g of NaCl is mixed with 100.0 g of water at 80.0 °C, how much will not dissolve?

   _____

9. If 15.0 g of KCl is added to 100.0 g of water at 30.0 °C, how much more must be added to saturate the solution?

   _____

10. If a saturated solution of KNO_3 at 20.0 °C is heated to 80.0 °C, how much more could be dissolved?

_____

11. If a saturated solution of KCl at 90.0 °C is cooled to 30.0 °C, how much of the solid will precipitate?

_____

12. How much NaCl will dissolve in 350.0 g of water at 70.0 °C?

_____

13. How much KCl will dissolve in 50.0 g of water at 50.0 °C?

_____

14. Classify as saturated or unsaturated a solution that contains 90.0 g of KNO_3 in 100.0 g of water at 60.0 °C.

_____

15. Classify as saturated or unsaturated a solution that contains 50.0 g of KCl in 100.0 g of water at 70.0 °C.

_____

16. What temperature is needed to dissolve twice as much KNO_3 as can be dissolved at 30.0 °C in 100.0 g of water?

_____

17. What is the solubility of O_2 in 1000.0 g of water at 24.0 °C?

_____

18. How much more soluble is O_2 in water at 16.0 °C than at 40.0 °C?

_____

19. At what temperature is the solubility of O_2 10.0 mg in 1000.0 g of water?

_____

20. What mass of O_2 can be dissolved in 2000.0 g of water at 30.0 °C?

_____

Unit 1

C.2 SUPPLEMENT: SOLUTION CONCENTRATION

Show work and units in all problems.

1. A solution of seawater contains 3.0 g salt in 97.0 g of water. What is the percent concentration?

2. What is the concentration of a sugar water solution that has 10.0 g of sugar in 40.0 g of water?

3. An acidic solution is made by dissolving 4.0 g of acetic acid in 35.0 g of water. What is the percent concentration of the solution?

4. A saturated solution of KNO_3 at 80 °C has 160.0 g of KNO_3 in 100.0 g of water. What is the percent concentration?

5. At 26 °C, water is saturated with oxygen when 0.008 g of oxygen gas is dissolved in 1000.0 g of water. What is the percent concentration?

6. A solution that has 0.0045 g of lead ions (Pb^{2+}) dissolved in 1000.0 g of water is considered dangerous to human health. Calculate the percent concentration.

7. A concentrated commercial solution of hydrochloric acid (HCl) is 37% by mass. If you had a solution of 550 g of the acid, how many grams of HCl would be in the solution?

8. Arsenic ions are a high risk to aquatic life at a concentration of 0.44 ppm. If a 500.0 g sample of water with this concentration is measured, how many grams of arsenic ions are present in the solution?

Unit 1

C.3 LABORATORY ACTIVITY: SOLUBILITY CURVES—PROCEDURE

Introduction

You have seen and used solubility curves earlier in this unit (pages 46–47). In this activity, you will collect experimental data to construct a solubility curve for succinic acid ($C_4H_6O_4$), a molecular compound. Before you proceed, think about how your knowledge of solubility can help you gather data to construct a solubility curve.

- How can the properties of a saturated solution be used?
- What temperatures can you investigate?
- How many times should you repeat the procedure to be sure of your results?

Discuss these questions with your partner or laboratory group. Your teacher will then discuss with the class how data will be gathered and will demonstrate safe use of the equipment that will be used.

Safety

Keep the following precautions in mind while performing this laboratory procedure.

- ◆ The succinic acid that you use is slightly toxic if ingested by mouth, so be sure to wash your hands thoroughly at the end of the laboratory.
- ◆ Never stir a liquid with a thermometer. Always use a stirring rod.
- ◆ Use insulated tongs or gloves to remove a hot beaker from a hot plate. Hot glass burns!
- ◆ Dispose of all wastes as directed by your teacher.

Procedure

1. To make a water bath, add approximately 300 mL water to a 400-mL beaker. Heat the beaker, with stirring, to either 45 °C, 55 °C, or 65 °C, as agreed to in your pre-lab class discussion. Ensure that the student team sharing your hot plate is investigating the same temperature. Carefully remove the beaker (using gloves or beaker tongs) when it reaches the desired temperature. NOTE: Do not allow the water-bath temperature to rise more than five degrees above the temperature that you have chosen. Return the beaker to the hot plate as needed to maintain the appropriate water-bath temperature.

2. Place between 4 g and 5 g succinic acid in each of two test tubes.

⚠ **CAUTION:** *Be careful not to spill any of the succinic acid. If you do, clean up and dispose of the succinic acid as directed by your teacher.* Add 20.0 mL distilled water to each test tube.

3. Place each test tube in the water bath and take turns stirring the succinic acid solution with a glass stirring rod every 30 seconds for 7 minutes. Each minute, place the thermometer in the test tube and monitor the temperature of the succinic acid solution, ensuring that it is within 2 °C of the temperature that you have chosen.

4. At the end of 7 minutes, carefully decant the clear liquid from each test tube into a separate, empty test tube, as demonstrated by your teacher.

5. Carefully pour the hot water from the beaker into the sink and fill the beaker with water and ice.

6. Place the two test tubes containing the clear liquid in the ice bath for 2 minutes. Stir the liquid in each test tube gently once or twice. Remove the test tubes from the ice water. Allow the test tubes to sit at room temperature for 5 minutes. Observe each test tube carefully during that time. Record your observations.

7. Tap the side of each test tube and swirl the liquid once or twice to cause the crystals to settle evenly on the bottom of the test tubes.

8. Measure the height of crystals collected (in millimeters, mm). Have your partner(s) measure the crystal sample height and compare your results. Report the average crystal height for your two test tubes to your teacher.

9. Rinse the succinic acid crystals from the test tubes into a collection beaker designated by your teacher. Make sure that your laboratory area is clean.

10. Wash your hands thoroughly before leaving the laboratory.

Data Analysis

1. Find the mean crystal height obtained by your class for each temperature reported.

2. Plot the mean crystal height in millimeters (y axis) versus the water temperature in degrees Celsius (x axis).

DATA TABLE

Temperature (°C)	Crystal Height #1 (mm)	Crystal Height #2 (mm)	Crystal Height #3 (mm)
45			
55			
65			

C.3 Laboratory Activity: Constructing a Solubility Curve

DATA TABLE—CLASSROOM DATA

Temp. (°C)	Group 1	Group 2	Group 3	Group 4	Group 5	Class Average
45						
55						
65						

Questions

1. _____

_____

_____

2. _____

_____

_____

3. _____

_____

_____

4. _____

_____

_____

5. _____

_____

_____

6. _____

_____

_____

Unit 1

C.8 LABORATORY ACTIVITY: SOLVENTS—PROCEDURE

Part I: Designing a Procedure for Investigating Solubility in Water

Depending on your teacher's advice, you will discuss, either with the whole class or with your laboratory partner, how you will go about testing the solubilities in room-temperature water of the substances that you listed in your laboratory notebook. (If you have performed solubility tests before, it may be useful to recall how you did that.) Then, with your partner, design a step-by-step investigation that will allow you to decide whether each solute is soluble (S), slightly soluble (SS), or insoluble (I) in room-temperature water. The questions below will help you design your procedure.

1. What particular observations will allow you to judge how well each solute dissolves in the polar solvent water? That is, how will you decide whether to classify a given solute as soluble, slightly soluble, or insoluble in water?

2. Which variables need to be controlled in the experiment? Why?

3. How should the solute and solvent be mixed—all at once or a little at a time? Why?

In designing your procedure, keep these concerns in mind:

♦ Hexane is highly flammable.

♦ Avoid direct contact of any solutes with your skin. Use wood splints to transfer solid samples; discard the splints when finished.

♦ Mix test-tube contents by gently tapping your forefinger against the side of the tube.

♦ Follow your teacher's directions for waste disposal.

♦ Record your results in a data table that indicates which solutes are soluble (S), slightly soluble (SS), or insoluble (I) in water.

When you and your partner have agreed on a written procedure, get it approved by your teacher. Then construct a data table for your results. You're ready for Part II!

Part II: Investigating Solubility in Water

Use your approved procedure to investigate the solubility in water of the substances in your lab notebook. Record the data in your data table.

It's clear that the task of determining what may have caused the fish kill can be simplified somewhat by focusing efforts on substances that will dissolve appreciably in water. However, in dealing with other solubility-based problems, chemists sometimes find it helpful to use a solvent other than water—solvents such as ethanol and hexane, for example.

Part III: Investigating Solubility in Ethanol and Lamp Oil

You and your partner will investigate the solubility of each of the solutes from Part II in the solvents ethanol and lamp oil. You should also test the solubility of water in ethanol and hexane. By gaining experiences with three liquid solvents—hexane, ethanol, and water—some general patterns regarding solubility behavior will emerge.

Can you use the same procedure you designed to be followed in Part II? If not, what parts of that procedure should be revised? In considering your Part III procedure, again keep these concerns and warnings in mind:

♦ Hexane is highly flammable.

♦ Avoid direct contact of any solutes with your skin. Use wood splints to transfer solid samples; discard the splints when finished.

♦ Mix test-tube contents by gently tapping your forefinger against the side of the tube.

♦ Follow your teacher's directions for waste disposal.

♦ Record your results in a data table that indicates which solutes are soluble (S), slightly soluble (SS), or insoluble (I) in either ethanol or lamp oil.

Have your proposed Part III procedure approved by your teacher. Before you start the laboratory work, test your interpretation of your Part II results by predicting (in writing on your Part III data table) what you think you will observe regarding solubility in ethanol and lamp oil. Then collect and record your data for both solvents.

Wash your hands thoroughly before leaving the laboratory.

C.8 LABORATORY ACTIVITY: SOLVENTS

Procedure for Part I and Part II

_____

_____

_____

_____

_____

_____

_____

_____

_____

_____

_____

_____

_____

_____

_____

_____

_____

_____

_____

_____

_____

_____

_____

_____

_____

_____ **Teacher Approval**

POSSIBLE DATA TABLE FOR PART II

Solute	Soluble (S), Insoluble (I), Slightly Soluble (SS) (in water)
Urea, $CO(NH_2)_2$	
Glucose, $C_6H_{12}O_6$	
Sodium chloride, NaCl	
Iodine, I_2	
Copper(II) sulfate, $CuSO_4$	
Calcium carbonate, $CaCO_3$	
Naphthalene, $C_{10}H_{18}$	
Ethanol, C_2H_5OH	
Hexane, C_6H_{14}	

Revision of procedure for Part III

_____

_____

_____

_____

_____

_____

_____

_____

_____

_____

_____

_____

_____ **Teacher Approval**

Name _____ Period _____ Date _____

_____
_____
_____
_____
_____
_____
_____
_____
_____
_____
_____

DATA TABLE FOR PART III

Solute	Prediction	Soluble (S), Insoluble (I), Slightly Soluble (SS)	
		Ethanol	Lamp Oil
Urea, $CO(NH_2)_2$			
Glucose, $C_6H_{12}O_6$			
Sodium chloride, NaCl			
Iodine, I_2			
Copper(II) sulfate, $CuSO_4$			
Calcium carbonate, $CaCO_3$			
Naphthalene, $C_{10}H_{18}$			
Ethanol, C_2H_5OH			
Hexane, C_6H_{14}			
Water, H_2O			

Unit 1

D.5 LABORATORY ACTIVITY: WATER SOFTENING—PROCEDURE

Getting Ready

In this laboratory activity, you will explore several ways of softening water. You will compare the effectiveness of three water treatments for removing calcium ions from a hard-water sample: sand filtration, treatment with Calgon, and treatment with an ion-exchange resin.

Procedure

1. Prepare the equipment as shown in Figure 33. Lower each funnel-stem tip into a test tube supported in a test-tube rack.

2. Fold four pieces of filter paper; insert one in each funnel. Number the funnels 1 to 4.

3. Funnel 1 should contain only the filter paper; it serves as the control. (Hard-water ions in solution cannot be removed by filter paper.) Fill Funnel 2 one-third full of sand. Fill Funnel 3 one-third full of Calgon. Fill Funnel 4 one-third full of ion-exchange resin.

4. Pour about 5 mL of hard water into each funnel. Do not pour any water over the top of the filter paper or between the filter paper and the funnel wall.

5. Collect the filtrates in the test tubes. NOTE: The Calgon filtrate may appear blue due to other additives in the softener. This will cause no problem. However, if the filtrate appears cloudy, some Calgon powder may have passed through the filter paper. In this case, use a new piece of filter paper and refilter the test-tube liquid.

6. Add 10 drops of sodium carbonate (Na_2CO_3) solution to each filtrate. Does a precipitate form? Record your observations. A cloudy precipitate indicates that the Ca^{2+} ion (a hard-water cation) was not removed.

7. Discard the test-tube solutions. Clean the test tubes thoroughly with tap water, and rinse with distilled water. Place the test tubes back in the test tube rack as in Step 2. Do not empty or clean the funnels; they are used in the next step.

8. Pour another 5-mL hard water sample through each funnel, collecting the filtrates in clean test tubes. Adjust test tube liquid heights, if necessary, to make all filtrate volumes equal.

9. Add one drop of Ivory brand liquid hand soap (not liquid detergent) to each test tube.

10. Stir each tube gently. Wipe the stirring rod before inserting it into another test tube.

11. Compare the cloudiness (**turbidity**) of the four soap solutions. Record your observations. The greater the turbidity, the greater the quantity of soap that dispersed. The quantity of dispersed soap determines the cleaning effectiveness of the solution.

12. Stopper each test tube; then shake each test tube vigorously, as demonstrated by your teacher. The more suds that form, the softer the water. Measure the height of suds in each tube and record your observations.

13. Wash hands thoroughly before leaving the laboratory.

D.5 LABORATORY ACTIVITY: WATER SOFTENING

Purpose

To explore several ways of softening water.

DATA TABLE

	Filter Paper	Filter Paper and Sand	Filter Paper and Calgon	Filter Paper and Ion-Exchange Resin
Reaction with sodium carbonate (Na_2CO_3)				
Degree of cloudiness (turbidity) with Ivory soap				
Height of suds				

Questions

1. _____

_____

_____

2. _____

_____

_____

3. _____

_____

_____

4. _____

_____

_____

PUTTING IT ALL TOGETHER
FISH KILL—FINDING THE SOLUTION

Some members of your class will moderate the Putting It All Together: "Fish Kill—Who Pays?" Others will represent the various special interest groups outlined in the book. **Please Note: This is a role-modeling activity! Get into your roles!** Use acting, costumes, banners, slogans, etc.

Each special interest group will decide on its position on the fish kill issue. Group members should decide who is responsible for the problem, what should be done to solve the problem, and finally, who should pay for it. Each group will have two minutes to present their case. Your group may choose among several possible approaches.

1. *Analytical:* This approach attempts to answer questions like what caused the problem, who caused the problem, and who should pay.

2. *Defensive:* This approach would propose "Why we should not pay!" Your arguments would explain why you are not at fault.

3. *Offensive:* This approach would suggest who is at fault. You place the blame on someone else for the fish kill, and you want that person or group to pay for the damages and correct the cause of the problem.

4. *Other:* Perhaps some combination of approaches or your own style would work better for you.

Each group must come up with a minimum of five arguments to support its position. The group must also prepare one question and one backup question to ask each of the other groups. If any other group asks your question, the examiner must be ready with the backup question. Each special interest group must fill out forms stating the position of their group, the arguments they will present, and the questions for the other groups.

Each special interest group must elect a "spokesperson." This person will address the Town Council and present arguments to the other citizens of Riverwood. Each group must also have an "examiner." This person will question the arguments of other groups.

After all groups have made their presentations and answered questions, each group will have one minute for rebuttal. A rebuttal is your group's chance to clarify your position, defend your actions, explain why you are not at fault, or suggest another group to be at fault. The spokesperson will address the rebuttal or, alternatively, a third person may be assigned to this task.

After all of the arguments and rebuttals, the town council members will adjourn to make a decision on the matter and suggest a course for further action. The town council should ensure that the course of action in their plan is within their authority to implement.

Finally, after the council renders its decision, each person will do a wrap-up activity. The reporters will write a newspaper article for the Riverwood News about the meeting and the decision. The article should be 200–300 words in length—enough to capture the facts, opinions, decisions, and reactions of the meeting. All others will write a letter to the editor of the Riverwood News, as a citizen of Riverwood, discussing the meeting and why you agree or disagree with the council's decision. The letter should be approximately 100 words in length and should come from the perspective of your special interest group.

PUTTING IT ALL TOGETHER
RIVERWOOD TOWN COUNCIL MEETING—GROUP WORKSHEET 1

Special Interest Group: _____

Spokesperson: _____

Examiner: _____

Other Group Member(s):

_____

_____

Position on the fish kill issue:

_____

_____

_____

_____

Argument 1:

_____

_____

Argument 2:

_____

_____

Argument 3:

_____

_____

Argument 4:

_____

_____

Argument 5:

_____

_____

PUTTING IT ALL TOGETHER
RIVERWOOD TOWN COUNCIL MEETING—GROUP WORKSHEET 2

Questions for Other Groups

Special Interest Group: _____

Town Council

1. _____

2. _____

Power Company

1. _____

2. _____

Agricultural Cooperative

1. _____

2. _____

Mining Company

1. _____

2. _____

Consulting Engineers

1. _____

2. _____

Chamber of Commerce

1. _____

2. _____

County Sanitation Commission

1. _____

2. _____

Taxpayers Association

1. _____

2. _____

PUTTING IT ALL TOGETHER
RIVERWOOD TOWN COUNCIL MEETING EVALUATION FORM

As you prepare for the group presentations, use the 10 topics listed below as guidelines. Your team's performance will be evaluated according to these standards. Then, as you listen to each special interest group, evaluate their presentations according to their performance. You may also evaluate your own group's performance according to these guidelines. Note that the grading system goes from 0 (poor performance) to 10 (excellent performance).

Special Interest Group Name _____

Team members: _____

Evaluator's Name:		Score
1. Organization: Team members were organized and well prepared. Assignments and forms were submitted on time.	Poor　　　　　　　Great 0　2　4　6　8　10	
2. Group Arguments: Students presented well thought-out arguments appropriate to their special interest group and based on the data. Arguments were based on facts, not personal attacks or sensationalism.	Poor　　　　　　　Great 0　2　4　6　8　10	
3. Questions for Other Groups: Questions were well conceived and designed to clarify the issue. Alternate questions were also prepared.	Poor　　　　　　　Great 0　2　4　6　8　10	
4. Quality of Rebuttals: Teams were prepared for questions and provided effective answers based on facts and data.	Poor　　　　　　　Great 0　2　4　6　8　10	
5. Application of Science Concepts: Students applied knowledge gained in the unit as they interpreted data and presented arguments. Answers to questions reflected appropriate application of science concepts.	Poor　　　　　　　Great 0　2　4　6　8　10	
6. Visual Aids: Students made appropriate use of audio-visual aids and presentation materials.	Poor　　　　　　　Great 0　2　4　6　8　10	
7. Group Involvement: All team members contributed to the presentation.	Poor　　　　　　　Great 0　2　4　6　8　10	
8. Presentation: Students effectively adopted their roles and presented information appropriate to their special interest group. Students used appropriate costumes and props to identify their group.	Poor　　　　　　　Great 0　2　4　6　8　10	
9. Article or Letter to Editor: Writing effectively summarized the Town Meeting and evaluated the decision from the perspective of the special interest group. The article or letter displayed appropriate clarity, grammar, and style.	Poor　　　　　　　Great 0　2　4　6　8　10	
10. Science, Technology, and Society: Students made appropriate connections between science and technology issues and the social issues in this exercise.	Poor　　　　　　　Great 0　2　4　6　8　10	
Total (100 points possible)		

Unit 2

A.3 LABORATORY ACTIVITY: METAL OR NONMETAL—PROCEDURE

Getting Ready

In this activity you will investigate some properties of seven elements and then decide whether each is a metal, a nonmetal, or a metalloid. You will examine the color, luster, and form of each element and will attempt to crush each sample with a hammer. You may also test the substance's ability to conduct electricity. (As an alternative, your teacher may demonstrate this test.) You will also determine the reactivity of each element with two solutions, hydrochloric acid, HCl(aq), and copper(II) chloride, $CuCl_2$(aq).

Procedure

1. *Appearance:* Observe and record the appearance of each element, including physical properties such as color, luster, and form. For the purposes of this activity, you can record the form as crystalline (like table salt), noncrystalline (like baking soda), or metallic (like iron).

2. *Conductivity:* If an electrical conductivity apparatus is available, use it to test each sample. **CAUTION:** *Avoid touching the bare electrode tips with your hands; they can deliver an uncomfortable electric shock.* Touch both electrodes to the element sample, but do not allow the electrodes to touch each other. If the lightbulb is connected to the electrodes lights, the sample has allowed electricity to flow through it. Such a material is called a **conductor.** If the bulb fails to light, the material is a **nonconductor.**

3. *Crushing:* Gently tap each element sample with a hammer. Based upon the results, decide whether the sample is **malleable** (flattens without shattering when struck) or **brittle** (shatters into pieces).

4. *Reactivity with acid:*

   a. Label seven wells of a clean well plate *a* to *g*.

   b. Place a sample of each element in its appropriate well. Each sample should either be a 1-cm length of wire or ribbon or 0.2–0.4 g of solid. You can estimate that as no larger than the size of a match head.

   c. Add 15–20 drops of 0.5 M HCl to each well that contains a sample. **CAUTION:** *0.5 M hydrochloric acid (HCl) can chemically attack skin if allowed to remain in contact for a long time.* If any hydrochloric acid accidently spills on you, ask a classmate to notify your teacher immediately. Wash the affected area immediately with tap water and continue for several minutes.

   d. Observe and record each result. The formation of gas bubbles indicates that a chemical reaction has occurred. A change in appearance of an element sample may also be an indication of a chemical reaction. Decide which elements reacted with the hydrochloric acid and which did not. Record your results.

e. Discard the well-plate contents as instructed by your teacher.

5. *Reactivity with copper(II) chloride:*

   a. Repeat Steps 4a and 4b (see above).

   b. Add 15–20 drops of 0.1 M copper(II) chloride ($CuCl_2$) to each well containing a sample.

   c. Observe each system for three to five minutes—changes may be slow. Decide which elements reacted with the copper(II) chloride and which did not. Recall the criteria you used in the acid test to determine if a reaction occurred. Record each result.

   d. Discard the well-plate contents as instructed by your teacher.

6. Wash your hands thoroughly before leaving the laboratory.

A.3 LABORATORY ACTIVITY: METAL OR NONMETAL

Purpose

The purpose of this activity is to decide if an element is a metal or nonmetal after investigating its chemical or physical properties.

Element	Appearance	Conductivity (optional)	Result of Crushing	Reaction with acid	Reaction with $CuCl_2(aq)$
a.					
b.					
c.					
d.					
e.					
f.					
g.					

Questions

1. | **Appearance** | Physical Property | Chemical Property |
   |---|---|---|
   | **Result of Crushing** | Physical Property | Chemical Property |
   | **Conductivity** | Physical Property | Chemical Property |
   | **Reaction with acid** | Physical Property | Chemical Property |
   | **Reaction with $CuCl_2(aq)$** | Physical Property | Chemical Property |

2. **Group 1**

_____

Group 2

_____

3. _____

_____

4.

Element	Metal	Nonmetal	Metalloid
a.			
b.			
c.			
d.			
e.			
f.			
g.			

Unit 2

A.6 SUPPLEMENT: THE PATTERN OF ATOMIC NUMBERS AND PREDICTING PROPERTIES

Using a Periodic Table, fill in the chart with the correct information.

Name	Symbol	Atomic Number	Mass Number	# of Protons	# of Neutrons	# of Electrons
Boron	B	5	11	5	6	5
Zinc						
	K					
		22				
			122			
				92		
		47			61	
	Fm					100
				78		
	Kr					
		86				

Given these known compounds—CO_2, KF, MgO, $CaCl_2$, Na_2O, Ga_2O_3, and $AlCl_3$—predict the formulas for the following combinations of elements.

1. Si and O _____

2. Ba and S _____

3. K and S _____

4. B and F _____

5. Li and Br _____

6. Sr and O _____

7. In and I _____

8. Ca and F _____

9. Al and S _____

10. H and O _____

Unit 2

B.2 LABORATORY ACTIVITY: CONVERTING COPPER—PROCEDURE

Getting Ready

You have seen many chemical reactions in your lifetime. Some, such as a fireworks display, are memorable. Others, such as the slow process of rusting, may not leave much of an impression. Have you ever stopped to think of what happens to the atoms involved in those reactions? Do the materials used to make fireworks still exist after the fireworks are lit and shot into the sky? What about the iron that turns into rust?

In this laboratory activity, you will start with a powdered sample of shiny, elemental copper, a metal you may be considering for your coin design. As you observe copper's chemical behavior in this activity, think about whether copper's properties would make it good to use for your coin design.

Procedure

1. Prepare a data table to record the masses you will determine in Steps 2 and 9.

2. Measure and record the mass of an empty, clean crucible. Add approximately 1 g copper powder to the crucible. Record the mass of the crucible with copper powder in it within the nearest 0.1 g. Find the actual mass of copper powder added by subtracting the mass of the empty crucible from the mass of the crucible with copper powder in it. Record the mass of copper powder.

3. Which properties of copper can be directly observed? Record your observations of the copper powder.

4. Set up the crucible, clay triangle, and burner as shown in Figure 13. The crucible lid should be slightly ajar.

5. Light the burner and adjust the flame so that it just touches the crucible's bottom.

6. Heat the crucible and its contents for two minutes. Remove the flame and then, using a glass stirring rod or spatula, break up the solid in the crucible to expose as much copper metal as possible. **CAUTION:** *Avoid touching the hot crucible.*

7. Continue heating for about 10 more minutes, removing the flame and breaking up the solid every 2 to 3 minutes.

8. When you have finished heating, extinguish the burner flame and allow the crucible and its contents to cool to room temperature. Answer Questions 1–2 while you are waiting.

9. After the crucible and its contents have cooled, find their mass. Use this value and the mass of the empty crucible to find the mass of the contents. Record these values in your data table.

10. Transfer your product to a clean 100-mL beaker. Label and store the beaker and product as directed by your teacher.

11. Put away the other materials. Wash your hands thoroughly before leaving the laboratory.

B.2 LABORATORY ACTIVITY: CONVERTING COPPER

Purpose

The purpose of this activity is to observe chemical and physical changes to copper.

DATA TABLE

Data	Mass (g)
a. Mass of empty, clean crucible	
b. Mass of crucible and copper	
c. Mass of copper	
d. Mass of crucible and copper after heating	
e. Mass of copper after heating	

Observation of copper powder

_____

_____

_____

Questions

1. _____

_____

_____

2. _____

_____

_____

3a. _____

b. _____

_____

B.3 SUPPLEMENT: RATES OF REACTION

Introduction

In this activity you will compare how the completion times of reactions vary when changing concentration, temperature, and surface area. Reactions that have long completion times have slow reaction rates. Reactions that have a small amount of time to complete are said to have a rapid rate of reaction. You may already have experience with reaction rates. You probably have noticed that the explosion of fireworks is instantaneous, while the rusting of a nail is slow. Why is that, since both reactions are oxidation reactions? At the completion of this activity you should be able to explain the possible causes for these differences in reaction rates.

Materials (for a class of 24 working in pairs)

12 pieces of chalk (typically 3-inch lengths)
12 50-mL beakers
84 test tubes (18 × 150 mm, holding 27 mL, 7 test tubes per group)
24 250-mL beakers
12 test-tube racks (space for 8 test tubes in each)
12 test-tube clamps
12 scissors
12 spoons or mortar and pestles
12 filter papers
2 L household vinegar (5% acetic acid, or 0.8M acetic acid solution)
12 100-mL graduated cylinders
12 30-mL graduated cylinders
hot water bath on hot plate (50 ºC)
1 L of distilled or deionized water
watch or wall clock with a second hand

Procedure

1. Using the procedure directed by your teacher, transfer 150 mL of vinegar (acetic acid) into a 250-mL beaker, and 50 mL distilled water into a 250-mL beaker. **Caution:** *Acetic acid is corrosive. If any splashes on your skin, wash it immediately with water, and inform your teacher.* Place each test tube in the test tube rack.

Part 1: Effect of Concentration on Reaction Rate

2. In the test tube rack, set up five empty test tubes. To vary the concentration, transfer 20 mL of vinegar from the beaker into one empty test tube, 15 mL in the next, 10 mL in the next, 5 mL in the next, and 0.0 mL in the last. Now add distilled water to make a total of 20 mL in each test tube. (In other words, add 0, 5 mL, 10 mL, 15 mL, and 20 mL, respectively, of distilled water to each of the five test tubes.)

3. Using the scissors cut two 0.5 cm chunks of the chalk, and then cut these in half. The goal is to have four chunks of chalk all the same size.

4. To prevent the acid from splashing, hold each test tube at an angle. Then slide one chunk of chalk into the solution. Record the start time and end time of each reaction.

Part 2: Effect of Particle Size or Surface Area on Reaction Rate

5. Cut off another 0.5 cm of chalk, and divide it in half. Using a spoon or mortar and pestle, grind up one of the chunk halves onto the filter paper.

6. Transfer 20 mL of vinegar into each of the 50-mL beakers.

7. In one beaker place the whole chalk chunk, and into the other beaker pour the powdered chalk. Record the start time and end time of each reaction.

Part 3: Effect of Temperature on Reaction Rate

8. Transfer 20 mL of vinegar into the last two clean test tubes.

9. Using a test-tube clamp, place one test tube in a hot water bath set at 50 °C. Allow about 10 minutes for the solution to reach the bath temperature; then return the test tube to the test tube rack.

10. Cut off another 0.5 cm chunk of chalk, and cut the chunk in half.

11. Drop one chunk half into each of the solutions. Record the start time and end time of each reaction, and any observations.

12. Clean and dispose of all materials as directed by your teacher.

Questions

1. The concentration of 5% acetic acid (vinegar) is about 0.8M. For the concentration test, calculate the concentration of each solution used (20 mL × tube conc. = mL vinegar used × 0.8M).

_____

_____

_____

2. Describe the effect of concentration on the rate of a reaction. Reminder: Reactions that have long completion times have slow reaction rates, and vice versa.

_____

_____

_____

3. Describe the effect of particle size or surface area on the rate of reaction.

_____

_____

_____

4. Describe the effect of temperature on the rate of reaction.

_____

_____

_____

5. Using the knowledge gained from this activity, explain why the reactions of fireworks are instantaneous and the rusting of a nail is slow.

_____

_____

_____

6. For each part of this experiment, identify the independent, dependent, and controlled variables.

_____

_____

_____

B.3 SUPPLEMENT 2: RATES OF REACTION

Introduction

In this activity you will devise a procedure to compare how the completion times of reactions vary when concentration, temperature, and surface area are changed. Reactions that have long completion times have slow reaction rates, and vice versa. You may already have experience with reaction rates. You probably have noticed that the explosion of fireworks is instantaneous, while the rusting of a nail is slow. Why is that, given that both reactions are oxidation reactions? At the completion of this activity you should be able to explain the possible causes for these differences in reaction rates.

Materials (for a class of 24 working in pairs)

24 cups (8-ounce cups for cold and hot drinks)
72 Alka Seltzer tablets
3 L water
refrigerator or ice bath

Procedure

Devise three safe procedures using Alka Seltzer tablets to evaluate chemical reaction rates for temperature, particle size or surface area, and concentration. Once your procedure is approved by your teacher, proceed with your testing.

Questions

1. Describe the effect of temperature on the rate of reaction.

   _____

   _____

2. Describe the effect of particle size or surface area on the rate of reaction.

   _____

   _____

   _____

3. Using the knowledge gained from this activity, explain why the reaction rates of the oxidation of gaseous gasoline might be different from its liquid form.

   _____

   _____

   _____

   _____

Unit 2

B.4 LABORATORY ACTIVITY: RELATIVE REACTIVITIES OF METALS—PROCEDURE

Getting Ready

You will observe and compare some chemical reactions of several metallic elements with solutions containing ions of other metals. In particular, you will investigate the reactions of three metals (copper, magnesium, and zinc) with four solutions. Each solution is an ionic compound that contains a certain metal ion—copper(II) nitrate, $Cu(NO_3)_2$; magnesium nitrate, $Mg(NO_3)_2$; zinc nitrate, $Zn(NO_3)_2$; and silver nitrate, $AgNO_3$.

Procedure

1. Devise an orderly procedure that will allow you to observe the reaction (if any) between each metal (except silver, Ag) and each of the four ionic solutions. You will conduct each reaction in a separate well of your wellplate; each well should contain five drops of 0.2 M solution and a small strip of metal. How many different combinations of metals and solutions will you need to observe? How will you arrange things so you can complete your observations efficiently, yet remain certain which metal and which solution are in each well?

2. Prepare a data table to help you organize the observations and results of the procedure you devise.

3. Obtain four 5-mm strips of each of the three metals to be tested. Clean the surface of each metal strip by rubbing it with sandpaper or emery paper. Record observations of each metal's appearance.

4. Complete your planned procedure, writing your observations in your data table. **CAUTION**: *Avoid letting the $AgNO_3$ solution come in contact with skin or clothing as it causes dark, permanent nonwashable stains.* If no reaction is observed, write NR in the table. Record the observed changes if a reaction does occur.

5. Dispose of your solid samples and wellplate solutions as directed by your instructor.

6. Wash your hands thoroughly before leaving the laboratory.

Proposed Procedure _____

_____

_____

_____

Proposed Procedure (continued)

_____

_____

_____

_____

_____ **Teacher Approval**

B.4 LABORATORY ACTIVITY:
RELATIVE REACTIVITIES OF METALS

Purpose

The purpose of this activity is to find out the relative reactivities of different metals.

DATA TABLE

Metal	$Cu(NO_3)_2$ Cu^{2+}	$Mg(NO_3)_2$ Mg^{2+}	$Zn(NO_3)_2$ Zn^{2+}	$AgNO_3$ Ag^+
Cu				
Mg				
Zn				

Questions

1. _____

2. _____

3. _____

4. _____

_____

5. _____

_____

6a. _____

b. _____

7a. _____

b. _____

8a. _____

b. _____

Unit 2

C.1 SUPPLEMENT: KEEPING TRACK OF ATOMS

Fill-in-the-Blanks

1. A chemical equation is balanced if there are _____ of each kind of _____ on both sides of the equation.

2. Before looking at equations, determine the number of atoms of each kind in each of the following:

   a. $CaCO_3$ = ____Ca, ____C, ____O

   b. $(NH_4)_2SO_4$ = ____N, ____H, ____S, ____O

   c. $3 H_2$ = ____H

   d. $4 Mg(OH)_2$ = ____Mg, ____O, ____H

   e. $Ba(NO_3)_2$ = ____Ba, ____N, ____O

3. Now look at the equations. Count the number of atoms of each kind on each side of the following and determine if the statement is a balanced equation.

   a. $2 Na + 2 H_2O \longrightarrow 2 NaOH + H_2$

Reactants		Products
_____	Na	_____
_____	H	_____
_____	O	_____

   Balanced? Yes _____ No _____

   b. $4 NH_3 + 6 NO \longrightarrow 5 N_2 + 6 H_2O$

Reactants		Products
_____	N	_____
_____	H	_____
_____	O	_____

   Balanced? Yes _____ No _____

4. For each of the following, show the number of each type of atom on each side of the reaction. Decide if the chemical equation is balanced or not.

   a. $NaCl + F_2 \longrightarrow NaF + Cl_2$

_____	Na	_____
_____	Cl	_____
_____	F	_____

   Balanced? Yes _____ No _____

8. ____ Fe_3O_4 + ____ H_2 $\longrightarrow$ ____ Fe + ____ H_2O

9. ____ HBr + ____ O_2 $\longrightarrow$ ____ Br_2 + ____ H_2O

10. ____ Al_2O_3 + ____ HCl $\longrightarrow$ ____ $AlCl_3$ + ____ H_2O

11. ____ NH_4OH + ____ $FeCl_3$ $\longrightarrow$ ____ $Fe(OH)_3$ + ____ NH_4Cl

12. ____ NH_3 + ____ O_2 $\longrightarrow$ ____ NO + ____ H_2O

13. ____ I_2 + ____ HNO_3 $\longrightarrow$ ____ HIO_3 + ____ NO_2 + ____ H_2O

14. ____ CaO + ____ P_2O_5 $\longrightarrow$ ____ $Ca_3(PO_4)_2$

15. ____ $NaOH$ + ____ $Al_2(SO_3)_3$ $\longrightarrow$ ____ Na_2SO_3 + ____ $Al(OH)_3$

Unit 2

C.3 SUPPLEMENT:
MOLAR MASS COMPUTATION AND CONVERSIONS

Molar Mass Computations

The **molar mass** of a substance is the mass of one mole (6.02×10^{23} units) of any substance. For the purpose of this course, we will use the nearest whole-number value of the atomic masses found on the Periodic Table.

    To find the molar mass of a substance, multiply the number of atoms of each element by the atomic mass of the element. Then add the masses of the various elements.

Example 1: What is the molar mass of iron(III) oxide, Fe_2O_3?

$$2\ Fe = 2 \times 56 = 112$$
$$\underline{3\ O\ \ = 3 \times 16 = \ \ 48}$$
$$FeO_3 \qquad\qquad 160\ \text{g/mol}$$

Example 2: What is the molar mass of magnesium hydroxide, $Mg(OH)_2$?

$$1\ Mg = 1 \times 24 = 24$$
$$2\ O\ \ = 2 \times 16 = 32$$
$$\underline{2\ H\ \ = 2 \times 1\ \ = \ \ 2}$$
$$Mg(OH)_2 \qquad = 58\ \text{g/mol}$$

Determine the molar mass of each substance.

   1. Nitrogen gas: N_2

   2. Sodium chloride (table salt): NaCl

   3. Sucrose (table sugar): $C_{12}H_{22}O_{11}$

   4. Chalcopyrite: $CuFeS_2$

   5. Malachite: $Cu_2CO_3(OH)_2$

Unit 2

C.4 SUPPLEMENT: PERCENT COMPOSITION OF MATERIALS AND CONSERVATION OF MASS

Example: What percent of iron(III) hydroxide, $Fe(OH)_3$, is oxygen?

Step 1: Find the molar mass of the compound.

$$
\begin{aligned}
1\ Fe &&=& 56\ g \\
3\ O &= 3 \times 16 &=& 48\ g \\
3\ H &= 3 \times 1 &=& 3\ g \\
\hline
&&=& 107\ g/mol
\end{aligned}
$$

Step 2: Find the percentage by dividing the part by the whole and multiplying by 100.

$$\frac{48\ g\ oxygen}{107\ g\ total} \times 100 = 45\%$$

1. What percent of magnesium bromide, $MgBr_2$, is magnesium?

2. What percent of glucose, $C_6H_{12}O_6$, is carbon?

3. What percent is zinc of $Zn_3(PO_4)_2$?

4. What percent of $AgNO_3$ is silver?

5. Which has the higher percent of aluminum, Al_2O_3 or $Al(NO_3)_3$?

Balance these equations: Mathematically show that the law of conservation of matter is upheld.

Example: $Zn + HCl \longrightarrow ZnCl_2 + H_2$

1	Zn	1
1	H	2
1	Cl	2

$Zn + 2HCl \longrightarrow ZnCl_2 + H_2$

Step 1: Balance the equation.

Step 2: Find molar masses, multiply by coefficient or subscripts and mathematically prove that reactants = products.

$$65 + 2(1 + 35) = 65 + 2(35) + 2(1)$$
$$65 + 72 = 135 + 2$$
$$137g = 137g$$

1. $Al + HCl \longrightarrow AlCl_3 + H_2$

2. $F_2 + NiI_3 \longrightarrow NiF_3 + I_2$

3. $Fe + O_2 \longrightarrow Fe_2O_3$

4. $NH_4OH + FeCl_3 \longrightarrow NH_4Cl + Fe(OH)_3$

Unit 2

C.5 LABORATORY ACTIVITY: RETRIEVING COPPER—PROCEDURE

Introduction

In Laboratory Activity B.2 (pages 116–117), you heated metallic copper, producing a black powder that you know to be copper(II) oxide (CuO). Because atoms are always conserved in chemical reactions, the original copper atoms must still exist. In this laboratory activity, you will attempt to recover those atoms of metallic copper.

Procedure

Part I: Separating copper(II) oxide (CuO) from the sample

Most likely, in Laboratory Activity B.2, not all of the original copper powder reacted with oxygen gas when you heated the copper in air. Some copper metal is still likely mixed in with the black copper(II) oxide. The first steps in this activity involve separating this mixture into copper and copper(II) oxide. To do this, you will add dilute hydrochloric acid (HCl) to the black powder. Copper metal does not react with hydrochloric acid, so it will remain as a solid. The black copper(II) oxide, however, reacts with hydrochloric acid to produce copper(II) chloride ($CuCl_2$) and water, as shown in this equation:

$$CuO(s) \quad + \quad 2\,HCl(aq) \quad \longrightarrow \quad CuCl_2(aq) \quad + \quad H_2O(l)$$
Copper(II) oxide   Hydrochloric acid   Copper(II) chloride   Water

1. Obtain your beaker containing the black powder from Laboratory Activity B.2. Look closely at its contents. Is the material uniform throughout? If not, why not?

2. Add 50 mL of 1 M HCl to the beaker containing the copper oxide mixture. Record your observations of the solution. **CAUTION:** *Hydrochloric acid may damage your skin. If some HCl does spill on your skin, ask another student to notify your teacher immediately. Begin rinsing the affected area with tap water immediately.*

3. Gently heat the mixture to about 40 °C on a hot plate. Heat for 15 minutes, stirring every few minutes with a glass rod.

4. Remove the beaker from the heat source. Allow any unreacted copper metal remaining to settle to the bottom of the beaker. Then slowly decant the liquid into another 100-mL beaker.

5. Set aside the second beaker (containing the liquid) for Step 9.

6. Wash the solid copper remaining in the first beaker several times by swirling it gently with distilled water. Discard the liquid washings as instructed by your teacher.

7. Measure and record the mass of a piece of filter paper. Transfer the solid copper to the paper, and allow it to dry overnight.

12. When finished, discard the used zinc chloride solution and the used zinc as directed by your teacher.

13. Wash your hands thoroughly before leaving the laboratory.

D.3 LABORATORY ACTIVITY: STRIKING IT RICH

Purpose

The purpose of this activity is to observe how the properties of a metal can be changed through chemical and heat treatments.

DATA TABLE

Condition	Appearance
Untreated penny	
Penny treated with Zn and $ZnCl_2$	
Penny treated with Zn and $ZnCl_2$ and heated in burner flame	

Questions

1a. _____

_____

_____

b. _____

_____

2. _____

_____

_____

3. _____

_____

4a. _____

_____

b. _____

_____

Unit 2
D.6 Laboratory Activity: COPPER PLATING

Introduction

As you just learned, plating a metal requires the application of direct current, the presence of metal ions, and a suitable anode, usually made of the metal to be plated. In this activity, you will plate copper onto a nail. You will need the following materials:

- Beaker, U-tube, or transparent plastic (Tygon) tubing
- Copper plating solution
- ⚠ **CAUTION:** *Copper plating solutions are hazardous and corrosive.*
- Iron or zinc nail
- Copper metal strips
- 9-V battery or power supply
- ⚠ **CAUTION:** *Always be careful when working with electricity, especially high-voltage power supplies.*
- Wire leads with alligator clips
- Voltmeter or CBL kit (optional)

Procedure

Set up a system to deposit copper onto a nail. Keep in mind which metal should act as the anode and which as the cathode. Think about how electrons will flow. Test your setup. Record your results. If nothing happens, have your teacher check your setup. When you are finished with the activity, be sure to follow your teacher's instructions for disposal of wastes. Wash your hands thoroughly before leaving the laboratory.

D.6 LABORATORY ACTIVITY: PLATING COPPER

Purpose
The purpose of this activity is to plate copper onto a nail.

Proposed Procedure

Draw a picture of the setup needed to plate copper onto the nail (*Hint:* see Figure 30, page 166). Label all parts. Then write a procedure to make and operate your apparatus.

8. Record the temperature at which the first drop of distillate enters the beaker. Then continue to record the temperature every 30 seconds. Continue to heat the flask and collect distillate until the temperature begins to rise again. At this point, replace the Distillate 1 beaker with the Distillate 2 beaker.

9. Continue heating and recording the temperature every 30 seconds until the second substance just begins to distill. Record the temperature at which the first drop of second distillate enters the beaker. Collect 1 to 2 mL of the second distillate. **CAUTION:** *Do not allow all of the liquid to boil from the flask.*

10. Turn off the heat and allow the apparatus to cool. While the apparatus is cooling, test the relative solubility of solid iodine (I_2) in Distillate 1 and Distillate 2 by adding a small amount of iodine to each beaker and stirring. Record your observations.

11. Disassemble and clean the distillation apparatus and dispose of your distillates as directed by your teacher.

12. Wash your hands thoroughly before leaving the laboratory.

A.2 LABORATORY ACTIVITY: SEPARATION BY DISTILLATION

Data Table

Observations of starting mixture being distilled _____

_____

_____

Distillation:	Time (sec)	Temperature (°C)
	0	_____
	30	_____
	60	_____
	90	_____
	120	_____
	150	_____
	180	_____
	210	_____
	240	_____

Temperature at which first drop of first distillate enters the beaker: _____

Temperature at which first drop of second distillate enters the beaker: _____

Name _____ Period _____ Date _____

Observations of distillates *before* iodine is added:

Distillate 1 _____

Distillate 2 _____

Observations of distillates *after* iodine is added:

Distillate 1 _____

Distillate 2 _____

Questions

1. Construct a graph with time on the *x*-axis and temperature on the *y*-axis.

2a. Temperature at which first substance distilled: _____

Temperature at which second substance distilled: _____

b. _____

_____

3. Distillate 1 _____ Distillate 2 _____

4a. Distillate 1 Mean _____ Mode _____

Distillate 2 Mean _____ Mode _____

b. _____

_____

5. _____

_____

6. _____

_____

7. _____

_____

_____

Unit 3

A.6 LABORATORY ACTIVITY: MODELING ALKANES—PROCEDURE

Introduction

In this activity you will assemble models of several simple hydrocarbons. You goal is to associate the three-dimensional shapes of these molecules with the names, formulas, and pictures used to represent them on paper.

Procedure

1. Assemble a model of methane (CH_4). Compare your model to the electron-dot and structural formulas on page 185. Note that the angles defined by bonds between atoms are not 90°, as you might think by looking at the structural formula. If you were to build a close-fitting box to surround a CH_4 molecule, the box would be shaped like a triangular pyramid, or a pyramid with a triangle as a base. A **tetrahedron** is the name given to this three-dimensional shape.

   Why would the shape of a methane molecule be tetrahedral? Assume that the four pairs of electrons in the bonds surrounding the carbon atom—all with negative charges—repel one another. That is, the electron pairs stay as far away from one another as possible, arranging themselves so that they point to the corners of a tetrahedron. The angle formed by each C—H bond is 109.5°, a value that has been verified with several experimental methods. The angles are not 90°, as they would be if methane were flat. Verify this shape for yourself by arranging the atoms in your model.

2. Compare your three-dimensional model of methane to the representation of a tetrahedral molecule in Figure 7 (page 186).

   a. How does the two-dimensional drawing (Figure 7) incorporate features that aid in visualizing the three-dimensional structure?

   b. Are there features of the two-dimensional figure that are difficult to translate into a three-dimensional structure? Explain.

   c. Translate your three-dimensional model into a two-dimensional drawing. Your drawing should convey the tetrahedral structure of methane.

3. Assemble models of a two-carbon and a three-carbon alkane molecule. Recall that each carbon atom in an alkane is bonded to four other atoms.

   a. How many hydrogen atoms are present in the two-carbon alkane?

   b. How many hydrogen atoms are present in the three-carbon alkane?

   c. Draw a ball-and-stick model, similar to the one in Figure 6 on page 185, of the three-carbon alkane.

4. a. Draw electron-dot and structural formulas for the two- and three-carbon alkanes.
   b. The molecular formulas of the first two alkanes are CH_4 and C_2H_6. What is the molecular formula of the third?

Examine your three-carbon alkane model and the structural formula you drew for it. Note that the middle carbon atom is attached to two hydrogen atoms, but the carbon atom at each end is attached to three hydrogen atoms. This molecule can be represented as $CH_3—CH_2—CH_3$, or $CH_3CH_2CH_3$. Formulas such as these provide convenient information about how atoms are arranged in molecules. For many purposes, such "condensed" formulas are more useful than molecular formulas such as C_3H_8.

Consider the formulas of the first few alkanes: CH_4, C_2H_6, and C_3H_8. Given the pattern represented by that series, try to predict the formula of the four-carbon alkane. If you answered C_4H_{10}, you are correct! The general molecular formula of all alkane molecules can be written as C_nH_{2n+2}, where n is the number of carbon atoms in the molecule. So even without assembling a model, you can predict the formula of a five-carbon alkane: If $n = 5$, then $2n + 2 = 12$, and the formula is C_5H_{12}.

5. Using the general alkane formula, predict molecular formulas for the rest of the first ten alkanes. After doing this, compare your molecular formulas with the formulas given in Figure 8 (page 187) to check your predictions.

The names of the first ten alkanes are also given in Figure 8. As you can see, each name is composed of a prefix, followed by -ane (designating an alkane). The prefix indicates the number of carbon atoms in the backbone carbon chain. To a chemist, meth- means one carbon atom, eth- means two, prop- means three, and but- means four. For alkanes with five to ten carbon atoms, the prefix is derived from Greek—pent- for five, hex- for six, and so on.

6. Write structural formulas for butane and pentane.

7. a. Name the alkanes with these condensed formulas:
   (i) $CH_3CH_2CH_2CH_2CH_2CH_2CH_3$
   (ii) $CH_3CH_2CH_2CH_2CH_2CH_2CH_2CH_2CH_3$
   b. Write molecular formulas for the two alkanes in Question 7a.

8. a. Write the formula of an alkane containing 25 carbon atoms.
   b. Did you write the molecular formula or the condensed formula of this compound? Why?

9. Name the alkane having a molar mass of
   a. 30 g/mol.
   b. 58 g/mol.
   c. 114 g/mol.

A.6 LABORATORY ACTIVITY: MODELING ALKANES

1. _____

2a. _____

_____

b. _____

_____

c.

3a. _____

b. _____

c.

4a.

2-carbon electron-dot formula	3-carbon electron-dot formula
2-carbon structural formula	3-carbon structural formula

b. _____

5. C_6 _____

C_7 _____

C_8 _____

C_9 _____

C_{10} _____

6.

Butane	Pentane

7a. (1) _____ (2) _____

b. (1) _____ (2) _____

8a. _____

b. _____

_____

9a. _____

b. _____

c. _____

Unit 3

A.7 LABORATORY ACTIVITY: ALKANES REVISITED—PROCEDURE

Introduction

In this activity you will use ball-and-stick molecular models to investigate such variations in alkane structures—variations that can lead to different properties.

Procedure

1. Assemble a ball-and-stick model of a molecule with the formula C_4H_{10}. Compare your model with those built by others. How many different arrangements of atoms in the C_4H_{10} molecule can be constructed?

Molecules that have identical molecular formulas but different arrangements of atoms are called **isomers**. By comparing models, convince yourself that there are only two isomers of C_4H_{10}. The formation of isomers helps to explain the very large number of compounds that contain carbon chains or rings.

2. a. Draw an electron-dot formula for each C_4H_{10} isomer.
   b. Write a structural formula for each C_4H_{10} isomer.

3. As you might expect, alkanes containing larger numbers of carbon atoms also have larger numbers of isomers. In fact, the number of different isomers increases rapidly as the number of carbon atoms increases. For example, chemists have identified three pentane (C_5H_{12}) isomers. Their structural formulas are shown in Figure 10, p. 190. Try building these and other models. Are other pentane isomers possible?

4. Now consider possible isomers of C_6H_{14}.
   a. Working with a partner, draw structural formulas for as many different C_6H_{14} isomers as possible. Compare your structures with those drawn by other groups.
   b. How many different C_6H_{14} isomers were found by your class?

5. Build models of one or more C_6H_{14} isomers, as assigned by your teacher.
   a. Compare the three-dimensional models built by your class with corresponding structures drawn on paper.
   b. Based on your examination of the three-dimensional models, how many different C_6H_{14} isomers are possible?

A.7 LABORATORY ACTIVITY: ALKANES REVISITED

1. _____

2.

a. electron-dot formula	electron-dot formula
b. structural formula	structural formula

3. _____

4a.

b. _____

5a. _____

_____

b. _____

Unit 3

A.7 SUPPLEMENT: NAMING BRANCHED ALKANES

Introduction

In this activity, you will either give the name for a molecule or write a condensed formula for the name.

Example 1: $CH_3-CH-CH_2-CH_3$
           $|$
           CH_2-CH_3

Answer: The longest chain is 5 carbons with a 1-carbon branch on the third carbon. 3-methylpentane

Example 2: 3-ethylhexane

Answer: $CH_3-CH_2-CH-CH_2-CH_2-CH_3$
          $|$
          CH_2-CH_3

ANSWERS

1. $CH_3-CH-CH_2-CH_2-CH_2-CH_2-CH_2-CH_3$
      $|$
      CH_3

1. _____

2. $\quad\quad\quad CH_2-CH_3$
   $\quad\quad\quad |$
   $CH_3-CH_2-CH-CH_2-CH_2-CH_2-CH_3$

2. _____

3. $CH_3-CH_2-CH-CH_3$
          $|$
          CH_3

3. _____

4. $CH_3-CH_2-CH_2-CH-CH_2-CH-CH_2-CH_2-CH_3$
              $|$ $|$
              CH_3 CH_3

4. _____

5. 2-methylpentane

5. _____

6. 3-ethyldecane

6. _____

7. 4-propylheptane

7. _____

8. 2,3-dimethyloctane

8. _____

● Unit 3

B.3 LABORATORY ACTIVITY: COMBUSTION—PROCEDURE

Introduction

In this activity, you will measure the heat of combustion of a candle (paraffin wax) and compare this quantity with known values for other hydrocarbons. You will also investigate relationships between the quantity of thermal energy released when a hydrocarbon burns and the structure of the hydrocarbon.

Procedure

1. Read the entire procedure and prepare a suitable data table to record all of your specified measurements—masses, volumes, and temperatures.

2. Hold a lighted match near the base of a candle so that some melted wax falls onto a 3 × 5 index card. Immediately push the base of the candle into the melted wax. Hold the candle there for a moment to fasten it to the card.

3. Determine the combined mass of the candle and index card. Record the value.

4. Carefully measure (to the nearest milliliter) about 100 mL of chilled water. (The chilled water, provided by your teacher, should be 10 to 15 °C colder than room temperature.) Pour the 100-mL sample of chilled water into an empty soft-drink can.

5. Set up the apparatus as shown in Figure 15 on page 204, but do not light the candle yet! Adjust the can so the top of the candlewick is about 2 cm from the bottom of the can.

6. Measure both room temperature and the water temperature to the nearest 0.1 °C. Record these values.

7. Place the candle under the can of water. Light the candle. As the water heats, stir it gently.

8. As the candle burns and becomes shorter, you may need to lower the can so the flame remains just below the bottom of the can. **CAUTION:** *Lower the can with great care.*

9. Continue heating until the temperature rises as far above room temperature as it was below room temperature at the start. (For example, if the water is 15 °C before heating and room temperature is 25 °C, you would heat the water to 35 °C, which is 10 °C higher than room temperature.)

10. When the desired temperature is reached, extinguish the candle flame.

11. Continue stirring the water until its temperature stops rising. Record the highest temperature reached by the water.

12. Determine the mass of the cooled candle and index card, including all wax drippings.

13. Wash your hands thoroughly before leaving the laboratory.

B.3 LABORATORY ACTIVITY: COMBUSTION

DATA TABLE

Mass of candle and index card		g
Volume of chilled water		mL
Room temperature		°C
Temperature of chilled water		°C
Highest temperature of water		°C
Mass of candle, index card, and wax drippings		g

CALCULATIONS

Mass of water heated		g
Rise in temperature of water		°C
Thermal energy used		J
Mass of paraffin wax burned		g
Heat of combustion of paraffin		J/g
Heat of combustion of paraffin		kJ/g

Questions

1. _____

_____

2. _____

_____

_____

3. _____

_____

4. _____

_____

5a. _____

_____

b. _____

_____

Unit 3

B.4 SUPPLEMENT: HEATS OF COMBUSTION

Solving Practical Problems Using Heats of Combustion

1. A gallon of gasoline can be thought of as 2660 g of octane.

   a. Find the number of kJ obtained by burning 1 gal of gasoline.

   _____

   b. Assume that your car gets 22 mi per gal. Find out how many kJ are used per mile.

   _____

   c. A well-tuned car may only use 25 percent of the energy burned for motion. How many kJ are "lost"?

   _____

2. The average person burns 12 000 kJ of food energy per day (2 800 000 cal). If the heat of combustion of sugar is 4000 cal/g, and the heat of combustion of fat is 9000 cal/g:

   a. How many grams of sugar must you eat to get a day's supply of energy?

   _____

   b. How many grams of fat must you eat to get a day's supply of energy?

   _____

   c. If your body ran on gasoline, how many grams of gasoline would you have to consume to get a day's supply of energy? How many gallons would this be?

   _____

3. Cars can run on grain alcohol (ethanol) instead of gasoline. A car that drives 15 000 mi per year and gets 20 mi/gal would require the alcohol made from 16 500 lbs of grain. An average person can survive if given 400 lbs of grain per year (355 medium-sized boxes of wheat flake cereal).

a. If an average acre of land produces 1900 lbs of grain per year, how many acres of grain must be grown to supply the average car with alcohol?

_____

b. How many people could the acres feed?

_____

c. How far can a car go on 1 lb of grain (one box of wheat flake cereal) that has been changed to alcohol?

_____

d. If there are 135 million cars in the United States today, how many acres of grain would be needed for alcohol?

_____

Unit 3

C.3 LABORATORY ACTIVITY: THE BUILDERS—PROCEDURE

Introduction

This activity, in which you will use models to simulate various arrangements of atoms, will help you to become more familiar with the alkenes and their polymers.

Procedure

Part 1: Alkenes

1. Examine the electron-dot and structural formulas for ethene, C_2H_4. Confirm that each atom has attained a filled outer shell of electrons.

$$
\begin{array}{ccc}
\overset{\displaystyle H \quad H}{H:\overset{..}{C}::\overset{..}{C}:H} & \overset{\displaystyle H \quad H}{H-C=C-H} & CH_2CH_2 \text{ or } C_2H_4 \\
\text{Electron-dot} & \text{Structural} & \text{Molecular formula} \\
\text{formula} & \text{formula} &
\end{array}
$$

2. Recall that the alkane general formula is C_nH_{2n+2}. Examine the molecular formulas of ethene (C_2H_4) and butene (C_4H_8). What general formula for alkenes do the molecular formulas suggest?

3. Assemble a model of an ethene molecule and a model of an ethane (C_2H_6) molecule. Compare the arrangements of atoms in the two models. Rotate the two carbon atoms in ethane about the single bond. Then try a similar rotation with ethene. What do you observe? Can you build a molecule in which you can perform a rotation about a double bond? Write a general rule to summarize your findings.

4. Build a model of butene (C_4H_8). Compare your model to those made by others. Remember that alkenes must contain a double bond.

   a. How many different arrangements of atoms in a C_4H_8 chain appear possible? Each arrangement represents a different substance—another example of isomers!
   b. Which structural formulas in Figure 19 correspond to models built by you or your classmates?

5. Does each of the following pairs represent isomers, or are they the same substance?
   a. $CH_2{=}CH{-}CH_2{-}CH_3$ or $CH_3{-}CH_2{-}CH{=}CH_2$
   b. $CH_2{=}\underset{\overset{|}{CH_3}}{C}{-}CH_3$ or $CH_3{-}\underset{\overset{\|}{CH_2}}{C}{-}CH_3$

6. How many isomers of propene (C_3H_6) are there? Support your answer with the appropriate structure(s).

7. Are these two structures isomers or the same substance? Explain.

$$CH_2-CH_2 \qquad CH_3-CH-CH_2$$
$$| \qquad | \qquad\qquad\qquad | \qquad |$$
$$CH_2 \quad CH_2 \qquad\qquad CH_2-CH_2$$
$$\diagdown \quad \diagup$$
$$CH_2$$

8. Based on your knowledge of molecules with single and double bonds between carbon atoms, assemble a model of a hydrocarbon molecule with a triple bond. Your completed model represents a member of the hydrocarbon series known as **alkynes**. Based on your understanding of how alkanes and alkenes are named, write structural formulas for

   a. ethyne, commonly called acetylene.
   b. 2-butyne.

9. Are alkynes saturated or unsaturated hydrocarbons? Explain.

Part 2: Compounds of Carbon, Hydrogen, and Singly Bonded Oxygen

10. Assemble as many different molecular models as possible using all nine of these atoms:

    2 carbon atoms (each forming four single bonds)
    6 hydrogen atoms (each forming a single bond)
    1 oxygen atom (forming two single bonds)

11. On paper draw a structural formula for each compound you have constructed, indicating how the nine atoms are connected. Compare your structures with those made by other classmates. After you are satisfied that all possible structures have been produced, answer these questions:

    a. How many distinct structures did you identify?
    b. What is the structural formula for each structure?
    c. Are all of these structures isomers? Explain.

12. Each compound you have identified possesses distinctly different physical and chemical properties.

    a. Recalling that "like dissolves like," which compound should be most soluble in water?
    b. Which compound should have the highest boiling point?

Part 3: Alkene-Based Polymers

In this part of the activity, you and your classmates will use models to simulate the formation of several addition polymers.

13. Build models of two ethene molecules.

14. Using information on page 217 as a guide, combine your two models into a dimer, or a two-monomer structure. What modifications in the monomer structure were necessary to accomplish this?

15. Combine your dimer with that of another lab team. Continue this process until your class has created a long-chain structure. Although your resulting molecular chain is not yet long enough to be regarded as a polymer, you have modeled the processes involved in creating a typical addition polymer.

16. Repeat Steps 13 and 14 for vinyl chloride and again for acrylonitrile. Then determine whether you can do the same for styrene. (See page 217 for the molecular structures for these substances.)

17. Assume the structures you built in Steps 15 and 16 became significantly larger. Give the name of each resulting polymer.

18. Are the chains you built linear or branched? How would this affect the properties of the polymer?

19. Make a polyethene chain that includes some cross-linking.

   a. Does this change the behavior of the model?
   b. How would the cross-linking change the properties of the polymer?

C.3 LABORATORY ACTIVITY: THE BUILDERS

Part 1

1. _____

2. _____
   _____

3. _____
   _____

4a. _____

 b. _____
   _____

5a. _____

 b. _____

6. _____

7. _____
   _____

8.

a.	b.

9. _____

_____

Part 2

10.

a.	b.	c.
d.	e.	f.

11a. _____

b.

i.	ii.	iii.
iv.	v.	vi.

c. _____

_____

12a. _____

b. _____

Part 3

13. Build models of two ethane molecules.

14. _____

_____

17. _____

_____

18. _____

_____

19a. _____

_____

b. _____

_____

Unit 3

C.4 SUPPLEMENT: BUILDERS

Compounds are classified according to the functional group that is present in their molecular structure. Find and circle the functional group in each molecule, and identify it as either a carboxylic acid, alcohol, ester, ether, or none of the above.

1. CH_3-CH_2-O-H

_____

2. 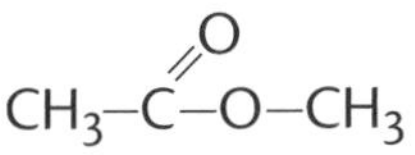 $CH_3-\overset{\displaystyle O}{\underset{}{C}}-O-CH_3$

_____

3. $H-O-\overset{\displaystyle O}{\underset{}{C}}-CH_3$

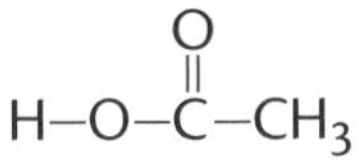

_____

4. 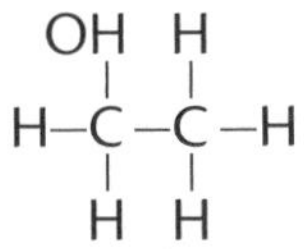 $\overset{OH}{\underset{H}{H-C}}-\overset{H}{\underset{H}{C}}-H$

_____

5. $CH_3-CH_2-O-CH_3$

_____

6. 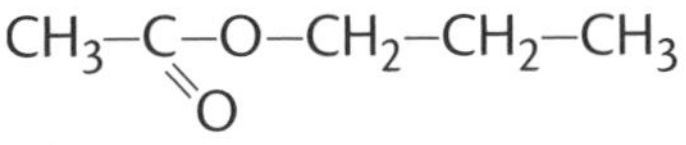 $CH_3-\overset{}{\underset{\displaystyle O}{C}}-O-CH_2-CH_2-CH_3$

_____

7. CH_3-O-CH_3

_____

8. 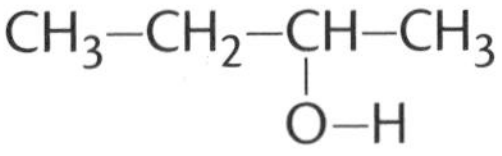 $CH_3-CH_2-\overset{}{\underset{O-H}{CH}}-CH_3$

_____

9. 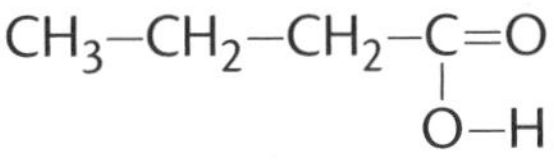 $CH_3-CH_2-CH_2-\overset{C=O}{\underset{O-H}{}}$

_____

10. $CH_3-CH_2-CH_2-CH_3$

_____

Unit 3

C.7 LABORATORY ACTIVITY: CONDENSATION—PROCEDURE

Introduction

In this activity you will produce several petrochemicals through the reaction of an organic acid (an acid derived from a hydrocarbon) with an alcohol. The esters you will produce have familiar, pleasing fragrances. Many perfumes contain esters; the characteristic aromas of many herbs and fruits arise from esters contained in the plants.

Procedure

1. Prepare a water bath by adding about 50 mL tap water to a 100-mL beaker. Place the beaker on a hot plate, and heat the water until it is near boiling.

2. Obtain a small, clean test tube. Place 5 drops methanol into the tube. Next add 0.1 g salicylic acid. Then add 2 drops concentrated sulfuric acid to the tube. **CAUTION:** *Concentrated sulfuric acid will cause burns to skin or fabric. Add the acid slowly and very carefully.*

3. As you dispense these reagents, note their odors. **CAUTION:** *Do not directly sniff any reagents—some may irritate or burn nasal passages.* Record any odors you happen to note.

4. Place the test tube in the water bath you prepared in Step 1.

5. Using test-tube tongs, move the test tube slowly in the water bath in a small horizontal circle. Keep the tube in the water, and do not spill the contents. Note any color changes. Continue heating for three minutes.

6. If you have not noticed an odor after three minutes, remove the test tube from the water bath, hold the test tube away from you with the tongs, and wave your hand across the top of the test tube to waft any vapors toward your nose. Record observations regarding the odor of the product. Compare your observations with those of other class members.

7. Repeat the procedure using 20 drops pentyl alcohol, 20 drops acetic acid, and 2 drops sulfuric acid.

8. Repeat the procedure using 20 drops octyl alcohol, 20 drops acetic acid, and 2 drops sulfuric acid.

9. Dispose of your products as directed by your teacher.

10. Wash your hands thoroughly before leaving the laboratory.

C.7 LABORATORY ACTIVITY: CONDENSATION

1. acid _____

    alcohol _____

2. _____

3. acid _____

    alcohol _____

4. _____

_____

_____

5a. _____

b. _____

c. _____

d. _____

C.7 FUNCTIONAL GROUPS

A functional group is a cluster of atoms that is found in various molecules. The group gives the molecule characteristic properties. Several groups may occur in one molecule giving it a mixture of properties.

Name	General Formula	Structural Formula	Name	Suffix (IUPAC)	Condensed Formula
Alkane	C_nH_{2n+2}	C–C	Ethane	–ane	CH_3CH_3
Alkene	C_nH_{2n}	C=C	Ethene	–ene	CH_2CH_2
Alkyne	C_nH_{2n-2}	C≡C	Ethyne	–yne	CHCH
Amine	RNH_2	C–C–N	Ethyl amine	amine	$C_2H_5NH_2$
Ether	ROR'	C–O–C	Dimethyl ether	---	CH_3OCH_3
Alcohol	ROH	C–C–OH	Ethanol (ethyl alcohol)	–ol	CH_3CH_2OH
Aldehyde	RCHO	C–C=O	Ethanal (ethyl aldehyde)	–al	CH_3CHO
Ketone	RCOR'	$\overset{\overset{\textstyle O}{\|\|}}{C-C-C-C}$	Butanone (methyl ethyl ketone)	–one	$CH_3COC_2H_5$
Carboxylic acid	RCOOH	C–C$\overset{O}{\underset{OH}{\diagup}}$	Ethanoic acid (acetic acid)	–oic acid	CH_3COOH
Aromatic	C_nH_n	⬡	Benzene	Benzene	C_6H_6
Ester	RCOOR'	C–C–C$\overset{O}{\underset{O-C}{\diagup}}$	Methyl propionate (propanoic acid methyl ester)	–oate	$C_2H_5COOCH_3$

R or **R'** stands for the rest of the molecule, other than the functional group, and signifies a group of atoms including at least one carbon atom bonded to the functional group. **R** and **R'** can be the same formula or different, depending on the compound.

Applications of Organic Functional Group Compounds

Simple hydrocarbons	Fuels, reactants
Alcohols	Solvents, reactants
Aldehydes and ketones	Flavorings, adhesives
Carboxylic acids	Preservatives, reactants
Aromatics	Medicines, plastics

C.7 GENERAL IUPAC RULES

General IUPAC Rules

1. Select the longest continuous chain of carbon atoms as the parent compound, and consider all alkyl groups attached to it as branched chains or substituents that have replaced hydrogen atoms of the parent hydrocarbon. If two chains of equal length are found, use the chain that has the larger number of substituents attached to it. The name of the alkane consists of the name of the parent compound prefixed by the names of the alkyl groups attached to it.

2. Number the carbon atoms in the parent carbon chain starting from the end closest to the first carbon that has an alkyl or other group substituted for a hydrogen atom. If the first substituent from each end is on the same-numbered carbon, go to the next substituent to determine which end of the chain to start numbering.

3. Name each alkyl group and designate its position on the parent carbon chain by a number (e.g., 2-methyl means a methyl group attached to carbon number 2).

4. When the same alkyl-group branch chain occurs more than once, indicate this repetition by a prefix (di-, tri-, tetra-, and so forth) written in front of the alkyl group name (e.g., dimethyl indicates two methyl groups). The numbers indicating the positions of these alkyl groups are separated by a comma and followed by a hyphen and are placed in front of the name (e.g., 2,3-dimethyl).

5. When several different alkyl groups are attached to the parent compound, list them in alphabetical order (e.g., ethyl before methyl in 3-ethyl-4-methyloctane). Prefixes are not included in alphabetical ordering (ethyl comes before dimethyl).

Some Examples of Naming Hydrocarbons with Functional groups
(IUPAC suffix listed first)

CH_3CH_3	ethane
CH_2CH_2	ethene
CH_3CCCH_3	butyne
$CH_3CH_2NH_2$	ethyl amine, or ethanamine
$CH_3CH_2OCH_3$	ethyl methyl ether (EME), or methoxyethane
CH_3CH_2OH	ethanol, ethyl alcohol, or ethyl hydroxide
$CH_3CH(OH)CH_3$	2-propanol, 2-propyl alcohol, or isopropyl alcohol
CH_3CHO	ethanal, or ethylaldehyde
$CH_3COCH_2CH_3$	2-butanone, or methyl ethyl ketone (MEK)
$CH_3CH_2CH_2COOH$	butanoic acid, butyric acid, or ethylacetic acid
C_6H_6	benzene, or cyclohexatriene
C_6H_5OH	phenol, phenyl hydroxide, or hydroxybenzene
$C_6H_5NH_2$	benzenamine, phenylamine, or aminobenzene

Unit 3

PIAT: PRESENTING & EVALUATING VEHICLE ADS

Use the following evaluation form to guide you as you develop the advertisements for an imaginary but plausible vehicle that uses *only* one specific type of fuel—electric, gasoline, hybrid gasoline-electric, or hydrogen in a fuel cell. You also may use the form to evaluate ads developed by others.

Specification	Scoring Guide	Points Earned (10 pt scale)	Multiplier (to 100 points or percent scale)	Score Earned
Time	1: A clear, concise message was conveyed within the "air time" limit. 0: Standards were not met.		x 10	
Scientific claims	3: All vehicle claims were accurate and explained, the vehicle was compared to petroleum vehicles, and fuel efficiency and emission projections were proposed for the design vehicle. 2: One of the claims was not covered. 1: Two of the claims were not covered satisfactorily. 0: None of the claims were adequately covered.		x 10	
Comfort, design, & safety features	3: The message presented a vehicle name, model, and special features to enhance consumer appeal and maintain safety parameters. 2: One of the designed features was not covered satisfactorily. 1: Two of the designed features were not covered satisfactorily. 0: None of the design standards were adequately covered.		x 10	
Presentation	3: The message was organized, visually stimulating, and motivating, while the script was interesting, clear, concise, and smoothly presented. 2: The message lacked one preparation feature. 1: The message lacked two preparation features. 0: None of the presentation standards were met.		x 10	
	Total (100 points possible)			

Unit 4

A.1 LABORATORY ACTIVITY: EXPLORING PROPERTIES OF GASES—PROCEDURE

Introduction

In this laboratory activity, you will perform a variety of experiments that illustrate some properties of air. Before you start this activity, carefully read through the procedure. Decide what you think will happen at each of the nine laboratory stations, and write down your predictions.

Procedure

Nine stations have been set up around the laboratory. At each station you will perform the experiment indicated. The experiments can be done in any order. For each station:

- Reread the procedure.
- Review your prediction.
- Perform the experiment.
- Record your observations.
- Restore the station to its original condition.

When you have completed your work at all the stations, answer the questions at the end of the activity.

Station 1

1. Inflate a balloon and tie the end.
2. Place the inflated balloon on a balance, using a piece of tape to hold it in place.
3. Record the mass.
4. Remove the balloon from the balance. Use a pin to gently puncture the balloon near the neck and release most of the gas contained in it.
5. Place the deflated balloon on a balance (with the tape still attached) and record its mass.

Station 2

1. With its open end facing downward, lower an empty drinking glass into a larger container of water.
2. With the open end still under water, slowly tilt the glass.

Station 3

Prediction _____

_____

_____

_____

Observation _____

_____

_____

Station 4

Prediction _____

Observation _____

_____

_____

Station 5

Prediction _____

_____

_____

Observation _____

_____

_____

Station 6

Prediction _____

_____

_____

Observation _____

_____

_____

Station 7

Prediction _____
_____
_____

Observation _____
_____
_____

Station 8

Prediction _____
_____

Observation _____
_____
_____

Station 9

Prediction _____
_____

Observation _____
_____
_____

Questions

1. _____
_____
_____
_____

2. _____
_____
_____
_____

3. _____

_____

_____

4a. _____

_____

_____

b. _____

_____

_____

c.

5a. _____

_____

_____

b. _____

_____

_____

6a. _____

_____

_____

b. _____

_____

_____

A.4 SUPPLEMENT: COMPRESSIBILITY OF MATERIALS

Introduction

In this activity you will investigate the shape and compressibility of a gas, two liquids, and a solid. To do this, you will put each type of substance into a closed syringe, note the shape of each, and attempt to compress each.

Background

Two of the characteristic properties of matter are shape and compressibility. The behavior of materials can change as their physical state—and consequently their characteristics—changes. Additionally, when defining shape, materials can be defined as fluids. A fluid is a substance that can flow and therefore conforms to the outline of its container. If a material does not flow, it is generally classified as a solid. Compressibility is a relative term that expresses how the volume of a substance changes as pressure changes. If the volume of the substance does not change, or changes insignificantly, the substance is relatively incompressible.

Time

15 minutes

Materials (for a class of 24 working in pairs)

12 syringes (between 10 to 30 mL, tip capped or sealed)
1 L of water (30 mL in a cup per group)
1 L of some other liquid designated by the teacher (30 mL in a cup per group)
1 lb of granulated table salt (30 mL of salt in a cup per group)
paper towels

Procedure

1. Prepare a chart for four materials: air, water, second liquid, and table salt. Provide space for observations of shape and compressibility.

2. Draw a syringe of air, and cap off the tip. Make a note of what shape the gas (air) is taking inside the syringe. As another way to demonstrate this, if your teacher permits it, light a match and draw the smoky air into the syringe, noting how the air suspended particles are distributed inside the syringe.

3. Record the initial volume. Holding the tip cap against a lab table or desk, try to compress the air in the syringe. *CAUTION:* **Do not compress the syringe without holding the cap firmly—otherwise, the cap could pop off with such force that it would become a hazardous projectile.** Note the change in volume.

4. Clean out the syringe, to ensure it is completely dry on the inside, and recap. Fill the syringe with salt. Note the shape that the salt takes in the syringe. Applying the same amount of pressure to the plunger as was used in the air test, repeat the compressibility test, and note the initial volume and change in volume.

5. Clean out and rinse the syringe thoroughly to remove all salt. Then fill the syringe with water. To ensure that there is no air in the syringe, point the tip upward, take off the cap, push up the plunger to remove the air, and recap the tip once all the air is purged. Note the shape the water takes inside the syringe, and repeat the compressibility test.

6. Repeat the shape and compressibility test with another liquid supplied by your teacher.

7. Clean the apparatus, disposing of the liquid in the syringe according to teacher directions, and wash your hands thoroughly.

Questions

1. Compare the compressibility of the four materials. If you could increase the pressure on the syringe to 1000 atmospheres, what results would you expect for the four materials?

_____

_____

2. Besides the shape of the syringe, is there any other factor that determines the shape a liquid takes?

_____

3. If the salt was replaced by a solid cylinder of copper that just fit into the syringe, what observations would you expect?

_____

_____

_____

4. What substances are fluids?

_____

5. What substances are relatively incompressible?

_____

6. Make a particle-level drawing of each substance in the syringe before and after compression. How did the structure and motion change for each? Compare your drawing and explanation with others in your class.

_____

_____

_____

_____

_____

Unit 4

A.5 SUPPLEMENT: BOYLE'S LAW

In this activity you will solve the following problems using Boyle's law. Show your setup for all problems.

1. A sample of 100.0 mL of oxygen has a pressure of 10.50 kPa. If the pressure is changed to 9.91 kPa, what is the new volume of the gas?

   _____

2. A flask containing 95.0 cm^3 of hydrogen was collected under a pressure of 731 mm of mercury. At what pressure would the volume be 70.0 cm^3?

   _____

3. A gas has a volume of 50.0 m^3. What volume would the gas occupy

   a. if the pressure is doubled? _____

   b. if the pressure is cut in half? _____

   c. if the pressure is tripled? _____

4. A scuba diver inflates a balloon at a depth of 99 ft (about 4 atm) to .25 ft^3. In the ascent will the balloon increase or decrease in volume?

   What will the volume be at the surface (1.0 atm)? _____

   _____

5. If the pressure of a gas is 3.0 atm and the volume of the gas doubles, what will the new pressure of the gas become?

   _____

6. A gas is confined in a cylinder with a movable piston at one end. When the volume of the cylinder is 760.0 cm^3 the pressure of the gas is 125.0 kPa. When the cylinder volume is reduced to 450.0 cm^3, what is the pressure in psi?

   _____

   _____

   _____

Unit 4

A.6 LABORATORY ACTIVITY: TEMPERATURE-VOLUME RELATIONSHIPS—PROCEDURE

Introduction

In this activity you will investigate how temperature changes of a gas sample influence its volume—assuming pressure remains unchanged. To do this, you will heat a thin glass tube containing a trapped air sample and record changes in air volume as the sample cools.

Procedure

1. Using two small rubber bands, fasten a capillary tube to the lower end of a thermometer. See Figure 15 (page 264). Place the open end of the tube closest to the thermometer bulb and 5–7 mm from its tip.

2. Immerse the tube and thermometer in a hot oil bath that has been prepared by your teacher. Be sure the entire capillary tube is immersed in oil. Wait for your tube and thermometer to reach the temperature of the oil (approximately 100 °C). Record the temperature of the bath.

3. After your tube and thermometer have reached constant temperature, lift them until only about one quarter of the capillary tube (open end down) is still in the oil bath. Pause for about 3 seconds to allow some oil to rise into the tube. Then quickly place the tube and thermometer on a paper towel (to avoid dripping) and carry them back to your desk. **CAUTION:** *Be careful not to touch the hot end of the thermometer or the drips of hot oil.*

4. Lay the tube and thermometer on a clean piece of paper towel on the desk. Make a reference line on the paper at the sealed end of the capillary tube. Also mark the upper end of the oil plug, as shown in Figure 15. Alongside this mark write the temperature corresponding to that air-column length.

5. As the temperature of the gas sample drops, make at least six marks representing the length of the air column trapped above the oil plug at various temperatures. Write the corresponding temperature next to each mark. Allow enough time so the temperature drops by 50–60 °C.

6. When the thermometer shows a steady temperature (near room temperature), make a final observation of length and temperature. Discard the tube and the rubber bands according to your teacher's instructions. Wipe the thermometer clean.

7. Measure the length (in centimeters) from each marked line to the mark for the sealed end of the tube. Record each length of the gas sample. Have your teacher check your data before you discard your paper towel.

8. Wash your hands thoroughly before leaving the laboratory.

A.6 LABORATORY ACTIVITY:
TEMPERATURE-VOLUME RELATIONSHIPS

Temperature of the hot oil bath _____°C

DATA TABLE

Temperature (°C)	Length of gas sample (cm)

Questions

1. _____

2a. _____

b. _____

_____

3. Renumber the temperature scale on your graph.

4. _____, _____

Unit 4

A.7 SUPPLEMENT: CHARLES' LAW

Use Charles' law to solve the following problems.
Show your setup on all problems.

1. If the temperature of a gas is 0.0 °C and the temperature is changed so that the gas volume doubles, what is the new temperature of the gas?

_____

2. A gas has a volume of 10.0 m^3 at standard temperature (273 K). What will the volume of the gas occupy if

a. the Kelvin temperature is doubled? _____

b. the Kelvin temperature is reduced to one-fourth of its original value?

_____

3. Suppose 500.0 mL of oxygen is collected at 25 °C. What will the volume be if the temperature is increased to 50 °C?

_____

4. A gas occupies a volume of 560.0 cm^3 at a temperature of 120 °C. In °C, to what temperature must the gas be lowered if it is to occupy 400.0 cm^3?

_____

5. Suppose 100.0 mL of nitrogen is collected at −10.0 °C. If the temperature of the gas is increased to 60.0 °C, how much will the gas increase in volume?

_____

6. A helium-filled balloon has a volume of 2.75 L at 20.0 °C. The volume of the balloon decreases to 2.46 L after it is taken outside on a winter day.

What is the outside temperature? _____

Unit 4

A.8 SUPPLEMENT: TEMPERATURE-PRESSURE RELATIONSHIPS

Solve the following problems. Show your set up on all problems.

1. An automobile tire has a pressure of 199 kPa at 20.0 °C. What will be the pressure after driving, if the tire temperature rises to 80.0 °C?

   _____

2. A cylinder of helium outside in the sun has a pressure of 2000.0 psi and a temperature of 51.0 °C. If the cylinder is taken indoors and cooled to 20.0 °C, what will its new pressure be?

   _____

3. To what temperature must a sample of nitrogen gas at 22.0 °C and 0.700 atm be heated so that its pressure increases to 1.25 atm?

   _____

4. The gaseous contents in an aerosol can are under a pressure of 3.00 atm at 25 °C. If the can pressure cannot exceed 4.00 atm without bursting, what is the highest Celsius temperature it can be exposed to?

   _____

5. An empty gasoline can, which still has gasoline vapors in it, has been stored in a garage since the summer when the temperature was 86.0 °F and the pressure was 29.92 in Hg. What would the pressure of the gas in the can be during the winter, when the temperature was 32.0 °F? Explain how this might effect the opening of the can.

   _____

   _____

   _____

6. A can of tennis balls are pressurized to 1.75 atm in a factory at 86.0 °F. If taken out on a wintry day of 10.0 °C, how much would the pressure change?

   _____

   _____

Unit 4

A.9 SUPPLEMENT: IDEAL GASES AND MOLAR VOLUME

In this activity you will solve these problems using your knowledge of molar volume and gas reactions.

1. 2.50 mol of neon gas would occupy what volume at 0.0 °C and 1 atm of pressure?

_____

2. 67.4 L of sulfur dioxide (SO_2) gas at 0.0 °C and 1 atm of pressure is equivalent to how many moles?

_____

3. Hydrogen chloride gas can be produced by a reaction between hydrogen gas and chlorine gas.

a. Write a balanced equation for this reaction.

_____

b. How many liters of hydrogen are needed to produce 1.75 L of hydrogen chloride?

_____

c. How many moles of chlorine would be needed to react with 8.65 mol of hydrogen?

_____

4. Tanks of gaseous propane (C_3H_8) are used for cooking and heating. When propane burns (using oxygen from the air), the products of the reaction are carbon dioxide and water vapor.

a. Write a balanced equation for this reaction.

_____

b. How many liters of oxygen would be needed to completely combust 0.350 L of propane?

_____

c. How many liters of water vapor would be produced by the reaction of 0.350 L of propane?

_____

5. A neon lightbulb (cylinder shaped, 2.60 cm × 120 cm) at 0.0 °C and 1 atm of pressure contains how many moles of gas?

_____

Unit 4

B.4 LABORATORY ACTIVITY: SPECIFIC HEAT CAPACITY—PROCEDURE

Introduction

As you have just learned, one characteristic property of a material is its specific heat capacity. In this activity, you will use this property to determine the identity of a metal sample.

Procedure

1. Half-fill a 250- or 400-mL beaker with water. Place the beaker on a hot plate and start heating.

2. Obtain a metal sample.

3. Determine and record the mass of the metal sample.

4. Place the metal sample into the water heating in the beaker. Allow the water to come to a boil, and keep it boiling for several minutes.

5. While the water is heating and boiling, obtain or set up a calorimeter. A simple version is shown in Figure 25 (page 284).

6. Accurately measure a volume of water that represents about 60–80% of the volume of the calorimeter. Record this value. Add the water to the calorimeter.

7. Determine and record the temperature of the water in the calorimeter.

8. Determine and record the temperature of the water boiling on the hot plate.

9. Once the water has boiled for several minutes, use tongs to remove the metal sample and place it in the calorimeter.

10. Stir the water in the calorimeter, and record its temperature every 30 seconds until it reaches a maximum value and starts to drop.

11. Wash your hands thoroughly before leaving the laboratory.

B.4 LABORATORY ACTIVITY: SPECIFIC HEAT CAPACITY

DATA TABLES

Mass of the metal sample	g
Volume of water in the calorimeter	mL
Beginning temperature of the water in the calorimeter	°C
Temperature of the water on the hot plate	°C

Time after emersion of metal in water	Water temperature
30 sec	
60 sec	
90 sec	
120 sec	
150 sec	
180 sec	
210 sec	
240 sec	
270 sec	
300 sec	
330 sec	
360 sec	
390 sec	
420 sec	
450 sec	
480 sec	

Calculations

1. Equation _____

Answer _____

2. Equation _____

Answer _____

Questions

1. _____

2. _____

_____

_____

_____

_____

_____

3a. _____

_____

_____

b. _____

_____

_____

4a. _____

_____

_____

b. _____

_____

_____

_____

Unit 4

B.6 LABORATORY ACTIVITY: CARBON DIOXIDE LEVELS—PROCEDURE

Introduction

In this activity you will estimate and compare the amounts of CO_2 in several air samples. To do this, the air will be bubbled through water that contains an indicator, bromthymol blue.

Procedure

Part 1: CO_2 in Normal Air

1. Pour 125 mL of distilled water into a 250-mL filter flask and add 10 drops of bromthymol blue. The solution should be blue. If it is not, add a drop of 0.1 M NaOH and gently swirl the flask. **CAUTION:** *Sodium hydroxide is corrosive. If any splashes on your skin, wash it immediately with water and inform your teacher.*

2. Pour 10 mL of solution prepared in Step 1 into a test tube labeled "control." Set this control aside for later comparisons.

3. Assemble the apparatus illustrated in Figure 28 (page 288).

4. Note and record the time. Then turn on the water tap until the aspirator pulls air through the flask. Mark or note the position of the faucet handle so you can run the aspirator at the same flow rate later in the experiment.

5. Let the aspirator run until the indicator solution turns yellow. Record the total time needed to reach the yellow color. Turn off the water. Remove the stopper from the flask.

6. Pour 10 mL of the used indicator solution from the flask into a second test tube labeled "Normal air." Stopper the tube. Compare the color of this sample with the control. Record your observations. Save the "Normal air" test tube.

Part 2: CO_2 from Combustion

7. Empty the filter flask and rinse it thoroughly with distilled water. Label a clean test tube "CO_2 combustion."

8. Place 125 mL of indicator solution, prepared according to Step 1, in the filter flask. Reassemble the apparatus as in Step 3.

9. Light a candle and position it so that the tip of the flame is just inside the base of the glass funnel attached to the flask.

10. Record the starting time. Then turn on the water tap to the position you established in Step 4. Run the aspirator until the indicator solution turns yellow. Record the time this takes. Turn off the water.

11. Pour 10 mL of the solution into a clean test tube labeled "CO_2 combustion." Stopper the tube. Compare its color with that of the "Normal air" and the "Control" solutions. Record your observations.

Part 3: CO_2 in Exhaled Air

12. Place 125 mL of indicator solution prepared according to Step 1 in a 250-mL Erlenmeyer flask.

13. Note and record the time. Then exhale your breath through a clean straw into the solution until the indicator color changes to yellow.

⚠ CAUTION: *Do not draw any solution into your mouth.* Record the total time it takes for the color to change.

14. Pour 10 mL of the solution into a clean test tube labeled "CO_2 breath." Stopper the tube. Compare its color with those of your other three solutions. Record your observations.

15. Dispose of the waste water solutions as directed by your teacher. Wash your hands thoroughly before leaving the laboratory.

B.6 LABORATORY ACTIVITY: CARBON DIOXIDE LEVELS

Data

Part 1: CO_2 in Normal Air

Beginning time _____

Time when color changed _____

Time it took for the color to change _____

Observations _____

_____

_____

Part 2: CO_2 from Combustion

Beginning time _____

Time when color changed _____

Time it took for the color to change _____

Observations _____

_____

_____

Part 3: CO₂ from Exhaled Air

Beginning time _____

Time when color changed _____

Time it took for the color to change _____

Observations _____

_____

_____

Questions

1. _____

_____

_____

_____

_____

2. _____

_____

_____

_____

3. _____

_____

_____

_____

4a. _____

_____

b. _____

_____

c. _____

_____

d. _____

_____

8. Place one drop of distilled water each on a fresh piece of red litmus paper, blue litmus paper, and pH paper. Record your observations.

9. Repeat Step 8, except use solution from test tube A in place of the distilled water. Record your observations.

10. Place a 1-cm length of magnesium ribbon in a separate clean, dry test tube. Add one pipet of the solution from test tube A. Observe the reaction for at least 3 minutes. Record your observations.

11. Add two small marble chips (calcium carbonate, $CaCO_3$) to the solution remaining in test tube A. Observe the marble chips for at least 3 minutes; record your observations.

12. Dispose of all remaining solutions as directed by your teacher.

13. Wash your hands thoroughly before leaving the laboratory.

C.2 LABORATORY ACTIVITY: MAKING ACID RAIN

Observations

Step 7 _____

_____

_____

Step 8 _____

_____

_____

Step 9 _____

_____

_____

Step 10 _____

_____

_____

Step 11 _____

_____

_____

_____

Questions

1. _____

2. _____

_____

_____

_____

_____

_____

_____

3. _____

_____

_____

_____

_____

_____

_____

_____

4. _____

_____

5. _____

_____

6. _____

_____

7a. _____

b. _____

Unit 4

C.8 LABORATORY ACTIVITY: BUFFERS—PROCEDURE

Introduction

In this activity, you will test a buffer solution. In particular, the results of adding acid and base to water, which serves as the control, will be compared to the results of adding acid and base to a buffered solution.

Procedure

1. Read the entire procedure and construct a data table suitable for collecting all relevant data.

2. Place a 24-well wellplate on a sheet of white paper. The white paper will help you more easily see any color changes during the activity. Add 20 drops distilled water to each of two wells. Add one drop universal indicator to each well. Note and record the colors of the resulting solutions in your data table.

3. Add 20 drops buffer solution to each of two other wells. This buffer is a solution containing 0.1 M sodium hydrogen phosphate (Na_2HPO_4) and 0.1 M sodium dihydrogen phosphate (NaH_2PO_4).

4. Add one drop universal indicator to each well containing buffer solution. Note and record the resulting color.

5. Add 5 drops 0.01 M NaOH to one of the wells containing distilled water and universal indicator. Note and record the color produced. **CAUTION:** ⚠ *The solutions in this experiment are corrosive. If you splash anything on your skin, wash it thoroughly with water and inform your teacher.*

6. Carefully counting each drop, slowly add 0.01 M NaOH to one of the wells containing the buffer solution and universal indicator until the color matches the color in the well from Step 5. Record the number of drops required.

7. Add 5 drops 0.01 M HCl to a second well containing distilled water and universal indicator. Record the color produced.

8. Carefully counting each drop, slowly add 0.01 M HCl to the second well containing the buffer solution and universal indicator until its color matches the color in the well from Step 7. Record the number of drops required.

9. Save the solutions until you have gathered all the information required in the data table.

10. Report your group's data to the class as instructed by your teacher.

11. Dispose of the solutions as directed by your teacher.

12. Wash your hands thoroughly before leaving the laboratory.

C.8 LABORATORY ACTIVITY: BUFFERS

DATA TABLE

Starting Material	Added	Initial Color	Approximate pH	Drops Required for Neutralization
Distilled water	Universal indicator			
Buffer solution	Universal indicator			
Distilled water and universal indicator	0.1 M NaOH			
Buffer solution and universal indicator	0.1 M NaOH			
Distilled water and universal indicator	0.1 M HCl			
Buffer solution and universal indicator	0.1 M HCl			

Questions

1. _____

_____

_____

2a. _____

_____

_____

b. _____

_____

_____

3. _____

Unit 4

D.6 LABORATORY ACTIVITY: CLEANSING AIR—PROCEDURE

Your teacher will demonstrate two control methods for air pollutants—electrostatic precipitation and wet scrubbing. Prepare a sketch of each laboratory setup as part of your note-taking for this activity.

Part 1: Electrostatic Precipitation Sketch

Observations

1. _____

_____

2. _____

_____

Part 2: Wet Scrubbing Sketch

Observations

3. _____

_____

_____

Questions

1. _____

_____

2. _____

_____

3. _____

_____

4. _____

_____

5. _____

_____

_____

6a. _____

_____

_____

b. _____

_____

_____

Unit 4

PUTTING IT ALL TOGETHER SUPPLEMENT
PROPOSAL FOR LUNAR HABITATION MODULE PROTOTYPE— AIR QUALITY MANAGEMENT PLAN

The Lunar Habitation Module Prototype (LHMP) Project is a proposal for the development of a self-contained human support system to be tested on Earth and duplicated on the lunar surface. Review the LHMP parameters and objectives listed on pages 246–247 in your text. Working with your team members, decide upon the design of the LHMP. As you complete your assignment, be sure to address the following aspects of your assignment.

Issue _____

_____

Key air quality issues _____

_____

_____

_____

_____

_____

Major emissions _____

_____

_____

_____

_____

_____

_____

Primary gases and their roles _____

_____

_____

_____

_____

_____

Monitoring emissions _____

_____

_____

_____

_____

_____

Options for managing emissions and protecting air quality _____

_____

_____

_____

_____

_____

_____

_____

Unit 4

PIAT PRESENTATION AND EVALUATION OF PROPOSAL

The following grading rubric provides a method for evaluating LHMP proposals. Discuss the specifications with your teacher. Remember to let the specifications guide your design proposals because your teacher will use the rubric to evaluate each group's proposal and presentation. Table headings should be centered. Multipliers and percentages should be in lower right bottom of the box.

Specification	Scoring Guide	Points Earned	Multiplier (to 100 points or percent scale)	Score Earned
Presentation	4: Message was very organized, visually stimulating, motivating, and smoothly presented, showing outstanding preparation. 3: Message was organized with an above-average presentation showing good preparation. 2: Message showed average group preparation and presentation. 1: Message was poorly organized and lacking in preparation. 0: None of the presentation standards were met.		x 5	
Identification of key air-quality issues and emissions	3: The proposal identified at least 5 issues and emissions. 2: The proposal identified at least 3 issues and emissions. 1: The proposal identified at least 1 issue and no emission. 0: The proposal identified no issue and emission.		x 5	
Description of roles of the primary gases involved	3: The proposal identified at least 3 roles. 2: The proposal identified at least 2 roles. 1: The proposal identified at least 1 role. 0: None of the criteria were met.		x 5	

(continued)

		x 5
Monitoring schemes	**3:** The proposal identified at least 2 emission problems and had a scheme that showed innovation, initiative, and a knowledge of the scheme's goals. **2:** The proposal identified at least 2 emission problems and had a scheme that showed good group effort. **1:** The proposal identified 1 emission and had a scheme, which was not well thought out or not feasible in nature. **0:** None of the criteria were met.	x 5
Recommended and alternative management options	**3:** The options were well researched and showed depth of knowledge. **2:** The options were researched and showed good effort by the group. **1:** The options were not well thought out and showed a lack of preparation. **0:** None of the criteria were met.	x 5
Formal written report	**4:** All requirements for the proposal were addressed, and the report included a scientific style of writing and appropriate graphics. **3:** All requirements for the proposal were addressed, and the report included a scientific style of writing. **2:** All requirements for the proposal were addressed, but the report was either poorly organized or had a number of grammatical errors. **1:** Not all of the requirements for the proposal were addressed; the report was incomplete. **0:** None of the criteria were met.	x 5
	Total	x 5

Unit 5

A.3 LABORATORY ACTIVITY:
FERTILIZER COMPONENTS—PROCEDURE

Introduction

In this laboratory activity you will test a fertilizer solution for six particular ions (three anions and three cations). In Part 1 you will perform tests on known solutions of those ions to become familiar with each confirming test. In Part 2 you will decide which ions are present in an unknown fertilizer solution. Read the complete procedure and prepare a suitable data table to record your observations.

Procedure

Part 1: Ion Tests

1. Prepare a warm-water bath for use in Step 6d. Pour about 30 mL of water into a 100-mL beaker. Place the beaker on a hot plate. The water must be warm, but should not boil. Control the heat accordingly.

2. Obtain a Beral pipet set containing each of six known ions—nitrate (NO_3^-), phosphate (PO_4^{3-}), sulfate (SO_4^{2-}), ammonium (NH_4^+), iron(III) (Fe^{3+}), and potassium (K^+). Record the color of each solution.

$BaCl_2$ Tests

Several ions you are studying in this activity can be identified first by their reaction with barium cations (Ba^{2+}) and then their behavior in the presence of acid.

3. a. Place a clean sheet of white paper under a multiple-well wellplate. Write the formulas of the six ions to be tested on the paper, locating each near a separate depression in the wellplate.

   b. Place 2 to 3 drops of each ion solution into its corresponding well.

   c. Test each sample solution individually by adding to it 1 or 2 drops of 0.1 M barium chloride ($BaCl_2$) solution. Use a new toothpick to mix each solution.

   d. Add 3 drops of 6 M hydrochloric acid (HCl) to each of the six wells containing $BaCl_2$ solution. **CAUTION:** *6 M HCl is corrosive. If any splashes on your skin, wash it off thoroughly with water and inform your teacher. Do not inhale HCl fumes.* Record your observations. Light-colored precipitates may be easier to observe if you temporarily remove the white paper beneath the wellplate. Again, record your observations.

   e. Clean and rinse the wellplate.

4. Dispose of all solutions as instructed by your teacher.

Brown-Ring Test

In the presence of nitrate ions (NO_3^-), mixing iron(II) ions (Fe^{2+}) and sulfuric acid (H_2SO_4) produces a distinctive result. This "brown-ring test" can be used to detect nitrate ions in a solution.

5. a. Place 8 drops of sodium nitrate ($NaNO_3$) solution in a small, clean test tube. Place 8 drops of distilled water in a second small, clean test tube, which will serve as the control.

   b. Add about 1 mL of iron(II) sulfate ($FeSO_4$) reagent to both test tubes. Gently mix each tube.

   c. Have your teacher carefully pour about 1 mL of concentrated sulfuric acid (H_2SO_4) along the inside of each test tube so the acid forms a second layer under the unagitated liquid already in the tube.

   ⚠ **CAUTION:** *Concentrated H_2SO_4 is a very strong, corrosive acid. If any contacts your skin, immediately wash affected areas with abundant running tap water and inform your teacher.*

   d. Allow the two test tubes to stand—without mixing—for 1 to 2 minutes.

   e. Observe any change that occurs at the interface between the two liquid layers. Record your observations.

NaOH and Litmus Tests

One or more of the three cations can be identified through observing their characteristic behavior in the presence of a strong base.

6. a. Add 4 drops of each cation test solution to three separate, clean test tubes.

   b. Moisten three pieces of red litmus paper with distilled water; place them on a watch glass.

   c. Add 10 drops of 3 M sodium hydroxide (NaOH) directly to the

   ⚠ solution in one test tube. **CAUTION:** *3 M NaOH is corrosive. If any splashes on your skin, wash it off thoroughly with water and inform your teacher.* Do not allow any NaOH solution to contact the test tube lip or inner wall. Immediately stick one of the three moistened red litmus paper strips from Step 6b onto the upper inside wall of the test tube. The strip must not contact the solution.

   d. Warm the test tube gently in the hot water bath for 1 minute. Note and record your observations after waiting about 30 seconds.

   e. Repeat Steps 6c and 6d for the other two test tubes.

Flame Test

Many metal ions can be identified by the characteristic color they emit when heated in a burner flame. It is common to use a flame test to identify potassium ions, for example.

7. a. Obtain a platinum or nichrome wire inserted into glass tubing or a cork stopper.

   b. Set up and light a burner. Adjust the flame to produce a light blue, steady inner cone and a more luminous, pale blue outer cone.

c. To clean the wire, place about 10 drops of 2 M hydrochloric acid (HCl) in a small test tube. **CAUTION:** *2 M HCl is corrosive. If any splashes on your skin, wash it off thoroughly with water and inform your teacher. Do not inhale HCl fumes.*

d. Dip the wire into the hydrochloric acid; then heat the wire tip in the burner flame. Position the wire in the outer "luminous" part of the flame, not in the center cone. As the wire heats to a bright red, the burner flame may become colored. See Figure 3. The colors are due to metallic cations held on the surface of the wire.

e. Continue dipping the wire into the acid solution and inserting the wire into the flame until there is little or no change in flame color as the wire is heated to redness.

f. Place 7 drops of potassium ion solution into a clean well in a wellplate. Dip the cool, cleaned wire into this solution. Then insert the wire into the flame. Note any change in flame color, the color intensity, and the time (in seconds) that the color is visible.

g. Repeat the potassium-ion flame test, this time observing the burner flame through cobalt-blue or didymium glass. Again, note the color, intensity, and duration of the color. Your partner can hold the wire in the flame while you observe through the colored glass. Then change places. Record all observations.

KSCN Test

In Unit 1 (page 37) you learned that when potassium thiocyanate (KSCN) is added to an aqueous solution containing iron(III) ions (Fe^{3+}), a deep red color appears due to formation of $[FeSCN]^{2+}$ cations. Appearance of this characteristic color confirms the presence of iron(III) in the solution.

8. a. Place 3 drops of iron(III)-containing solution into a well in a wellplate.

b. Add 1 drop of 0.1 M potassium thiocyanate (KSCN) solution to the well. Record your observations.

c. Clean and rinse the wellplate.

Part 2: Tests on Fertilizer Solution

1. Obtain a Beral pipet containing an unknown fertilizer solution. Record the code number of the solution. Your unknown solution contains one of the anions and one of the cations you tested in Part 1. Observe and record the color of the unknown solution.

2. Conduct suitable laboratory tests on the unknown solution until you are confident that you have identified which anion and cation from Part 1 are present. Record all observations and conclusions. Repeat a particular test if you wish to confirm your observations.

3. Dispose of all solutions used in this activity as directed by your teacher.

4. Wash your hands thoroughly before leaving the laboratory.

A.3 LABORATORY ACTIVITY: FERTILIZER COMPONENTS

Purpose

The purpose of this activity is to identify six ions and test a fertilizer solution for those ions.

Parts 1 (Ion Tests) and 2 (Tests on Fertilizer Solution)

Record test results for Parts 1 and 2 in the tables on page 141. Record answers to questions for both parts on the lines provided below.

Questions

1. _____

_____

_____

2. _____

_____

_____

3. _____

_____

_____

Name _____ Period _____ Date _____

Part 2: Tests on Fertilizer Solution

Ion Tests	Known Anions			Known Cations			Unknown # ___
	Nitrate (NO_3^-)	Phosphate (PO_4^{3-})	Sulfate (SO_4^{2-})	Ammonium (NH_4^+)	Iron(III) (Fe^{3+})	Potassium (K^+)	
Color							
$BaCl_2$							
$BaCl_2$ and HCl							
Brown-ring test		X	X	X	X	X	
NaOH and litmus test	X	X	X				
Flame test	X	X	X	X	X		
Flame test and cobalt glass	X	X	X	X	X		
KSCN test	X	X	X	X		X	

X = test not done

Ions	Ions Present in Unknown	Which Test(s) Confirmed the Ion?
Nitrate (NO_3^-)		
Phosphate (PO_4^{3-})		
Sulfate (SO_4^{2-})		
Ammonium (NH_4^+)		
Iron(III) (Fe^{3+})		
Potassium (K^+)		

Unit 5

A.5 LABORATORY ACTIVITY: PHOSPHATE—PROCEDURE

Introduction

Fertilizers can be evaluated in part by the percentages (by mass) of essential nutrients contained in them. In this activity you will analyze a fertilizer solution to determine the mass and percent of phosphate.

Procedure

1. Label five clean test tubes as follows: 10 ppm, 7.5 ppm, 5.0 ppm, 2.5 ppm, and x ppm.

2. To prepare a water solution of the unknown fertilizer, place a 0.50-g sample of the solid fertilizer in a 400-mL beaker. Add 250 mL of distilled water and stir until the sample is completely dissolved.

3. Measure out and retain 1/50 of your total fertilizer solution (that is, 5.0 mL of the original 250 mL). Discard the remaining volume of original solution as directed by your teacher.

4. Pour the 5.0-mL sample of solution into a clean 400-mL beaker. Then add enough distilled water to bring the total volume of solution in the beaker to 250 mL. Stir thoroughly.

5. Pour 20 mL of the diluted solution into the test tube labeled "x ppm." Discard the rest of the fertilizer solution remaining in the 400-mL beaker as directed by your teacher.

6. Your teacher has already prepared a supply of 10-ppm phosphate ion standard solution. Place 20 mL of that standard solution in the test tube labeled "10 ppm."

7. Given supplies of 10-ppm solution and distilled water, decide what volumes of 10-ppm solution and distilled water should be measured and mixed to prepare 20-mL samples, respectively, of 7.5-ppm, 5.0-ppm, and 2.5-ppm phosphate solutions. Write your plan in your laboratory notebook.

8. Ask your teacher to check your solution-preparation plan. After receiving your teacher's approval, prepare the three solutions. Pour each standard solution into its appropriately labeled test tube.

9. Add 2.0 mL of ammonium molybdate–sulfuric acid reagent to each of the four phosphate standards and also to the unknown solution.

10. Add a few crystals of ascorbic acid (no more than the volume of a pencil eraser tip) to each tube. Stir to dissolve. Rinse and dry the stirring rod after mixing each tube.

11. Place a 400-mL beaker half-full of water on a hot plate. Carefully place your five test tubes into the beaker. Heat the water bath until a blue color develops in the 2.5-ppm solution. Do not boil the water. Turn off the hot plate.

12. Using a test-tube holder, remove the test tubes from the water bath and place them in numerical order in a test-tube rack.

13. Compare the color intensity of the unknown solution ("x ppm") with the intensities of the four color-standard solutions. Place the unknown-solution test tube between the two tubes containing standard solutions with the closest-matching color intensities.

14. Estimate the concentration (ppm) of your unknown phosphate solution from the known color standards. For example, if the unknown solution color falls between the 7.5-ppm and 5.0-ppm color standards, you might decide to call it "6 ppm," or "6 g PO_4^{3-} per 10^6 g solution." Record the estimated phosphate ion concentration in ppm.

15. Discard your test-tube solutions as directed by your teacher.

16. Wash your hands thoroughly before leaving the laboratory.

A.5 LABORATORY ACTIVITY: PHOSPHATES

Purpose

The purpose of this activity is to use colorimetry to detect the concentration of phosphate ions in solution.

Solution Preparation Plan

_____

_____

_____

_____

_____

_____

_____ **Teacher Approval**

DATA TABLE

Concentration	Color Intensity of Solution
7.5 ppm	
5.0 ppm	
2.5 ppm	

Unknown ion solution estimated concentration _____ ppm

Calculations

1. Math setup to find mass of phosphate:

Mass of phosphate (in grams) _____

2. Math setup to find percent phosphate:

Percent phosphate by mass _____

Questions

1. _____

2. _____

_____

_____

_____

3. _____

_____

_____

4. _____

_____

5a. _____

_____

b. _____

_____

6. _____

_____

_____

Unit 5

B.2 LABORATORY ACTIVITY: LeCHATELIER'S PRINCIPLE—PROCEDURE

Introduction

In this activity you will take a system at equilibrium and use what you have learned about LeChatelier's Principle to investigate the effect of changes in concentration and temperature on a system at equilibrium. The chemical system you will investigate is described by this equation:

$$\text{Heat} + [Co(H_2O_6)]^{2+}(aq) + 4\ Cl^-(aq) \rightleftharpoons [CoCl_4]^{2-}(aq) + 6\ H_2O(l)$$

Procedure

1. Add 20 drops of 0.1 M cobalt(II) chloride, $CoCl_2$, solution to a clean, dry test tube. Record the color.

2. Add 7 drops of 0.1 M silver nitrate, $AgNO_3$, solution. **CAUTION:** *AgNO_3* *solution can stain skin and clothing. Handle it carefully.* Gently swirl the tube to ensure good mixing. Record the color.

3. Heat the tube in a hot water bath for 30 seconds. Record the color.

4. Remove the tube from the hot water bath. Add approximately 0.3 g sodium chloride, NaCl. Gently swirl the tube. Heat the solution for 30 seconds. Record the color.

5. Place the test tube in a beaker containing ice water for 30 seconds. Record the color.

6. Reheat the test tube in the hot water bath. Record the color.

7. Dispose of the mixture in the test tube as directed by your teacher.

8. Wash your hands thoroughly before leaving the laboratory.

Questions

In answering these questions, refer to your observations and to the equilibrium expression that appears in the introduction.

1. Which reaction (forward or reverse) was favored by cooling the solution?

2. Which reaction was favored by adding more chloride ions?

3. What is the identity of the white precipitate that formed in Step 2? (*Hint:* Refer to Unit 1, page 37.)

4. Why did adding $AgNO_3$ solution affect the equilibrium, even though neither Ag^+ ions nor NO_3^- ions appear in the equilibrium equation?

5. Why did the color change after heating in Step 4, but not in Step 3?

6. Which complex ion is pink—$[CoCl_4]^{2-}$ or $[Co(H_2O)_6]^{2+}$? Which complex ion is blue? Explain how you decided.

B.2 LABORATORY ACTIVITY: LeCHATELIER'S PRINCIPLE

DATA TABLE

Procedure Number	Color
1. $CoCl_2$	
2. $CoCl_2$ and $AgNO_3$	
3. $CoCl_2$ and $AgNO_3$, heated	
4. Add NaCl and heat again	
5. After ice bath	
6. Reheating	

Questions

1. _____

2. _____

3. _____

4. _____

_____

_____

5. _____

_____

_____

6. _____

_____

_____

Unit 5

B.7 MAKING DECISIONS SUPPLEMENT— WHAT DOES RIVERWOOD WANT?

Begin to prepare for your participation in the upcoming town meeting.

1. _____
_____
_____
_____

2.

Expectations	Mandatory or Desirable
1.	
2.	
3.	
4.	
5.	
6.	
7.	

3. _____

_____

_____

_____

Unit 5

C.1 SUPPLEMENT: ELECTROCHEMICAL CHANGES

1. When copper, Cu, is placed in a concentrated nitric acid, HNO_3, solution, vigorous bubbling takes place as a brown gas is produced. The copper disappears and the solution changes from colorless to bluish-green. The brown gas is nitrogen dioxide, NO_2, and the bluish-green solution is due to the formation of copper(II) ions, Cu^{2+}. Write a balanced chemical equation for this reaction. Which element (copper or nitrogen) is oxidized? Which element is reduced?

   _____

   _____

2. Putting iron into hydrochloric acid (aqueous hydrogen chloride) produces iron(II) chloride and hydrogen gas (H_2).

   a. Using Figure 18 on page 33, write the equation using chemical symbols and balance the equation.

   _____

   b. Does this equation represent an oxidation-reduction reaction?

   _____

   c. If so, identify the element oxidized and the element reduced. If not, explain why.

   _____

   _____

   _____

3. Sodium chloride is produced when sodium hydroxide reacts with hydrochloric acid. The other product is water.

   a. Using Figure 18 on page 33, write the equation using chemical symbols and balance the equation.

   _____

   b. Does this equation represent an oxidation-reduction reaction?

   _____

c. If so, identify the element oxidized and the element reduced. If not, explain why.

_____

_____

_____

4. In an experiment in Unit 2, you discovered that when magnesium was placed a solution of copper(II) ions, copper metal could be recovered. Using the information on oxidation-reduction reactions and electro-negativity, explain why this can happen.

_____

_____

_____

Unit 5

C.2 LABORATORY ACTIVITY: VOLTAIC CELLS—PROCEDURE

Introduction

In this activity you will construct several voltaic cells and measure and compare their electrical potentials. You will also explore factors that may help determine the electrical potential generated by a particular voltaic cell.

Procedure

Part I: Constructing a Voltaic Cell

1. Add 1 mL of 0.1 M $Cu(NO_3)_2$ to one well in a wellplate.

2. Add 1 mL of 0.1 M $Zn(NO_3)_2$ to an adjacent well.

3. Add a Cu strip (electrode) to the well containing $Cu(NO_3)_2$ solution.

4. Add a Zn strip (electrode) to the well containing $Zn(NO_3)_2$ solution.

5. Drape a small strip of filter paper saturated with KNO_3 solution between the wells containing the two solutions. Ensure that the filter paper strip is immersed in both solutions. Do not allow the metal strips to touch each other.

6. Obtain a voltmeter and two electrical wires with alligator clips. Attach one end of each wire to a separate voltmeter terminal.

7. Attach one wire from the voltmeter to the copper electrode. Lightly touch the second wire to the zinc electrode. If the needle deflects in the direction of a positive potential or if the digital readout is positive, attach the clip to the zinc metal. If the needle deflects in a negative direction (or the readout value is negative), reverse the clip connections to the electrodes. Figure 16 depicts a completed cell.

8. Record the electrical potential indicated by the voltmeter.

Part II: Measuring Electrical Potentials

9. Using what you learned in Part I about constructing a voltaic cell, construct additional cells, and measure electrical potentials for the voltaic cells composed of all possible pairs of half-cells listed below. Record your data.

   a. 0.1 M $Cu(NO_3)_2$ and Cu strip

   b. 0.1 M $Zn(NO_3)_2$ and Zn strip

   c. 0.1 M $Mg(NO_3)_2$ and Mg strip

   d. 0.1 M $Fe(NO_3)_2$ and Fe "strip" (nail)

Unit 5

C.3 BUILDING SKILLS SUPPLEMENT: GETTING A CHARGE FROM ELECTROCHEMISTRY

Half-reactions help a chemist understand the electron transfer during an oxidation-reduction reaction. Here are two examples: copper(II) ions to copper, and chloride ions to chlorine.

a. Cu^{2+} to Cu $Cu^{2+} + 2e^- \rightarrow Cu$ This is reduction.

b. Cl^- to Cl $Cl^- \rightarrow Cl + 1e^-$ This is oxidation.

To balance these half-reactions, the second equation would have to be doubled in order to equalize the electrons. The balanced equation then would be

$Cu^{2+} + 2Cl^- \rightarrow Cu + 2Cl$ (which in reality should be Cl_2)

1. Equations for the two half-reactions taking place in an electrochemical cell are

$Cr \rightarrow Cr^{3+} + 3e^-$ and $Pb^{2+} + 2e^- \rightarrow Pb$

Write a balanced equation for this reaction.

_____

_____

2. Two half-cells, one containing Ag^+ and Ag and the other containing Fe and Fe^{3+}, are hooked up to form a voltaic cell. Write the half-reactions and the balanced equation for this cell.

_____

_____

3. In the reaction between potassium (K) and water, potassium is oxidized and loses one electron. The hydrogen atoms in water are reduced, each gaining one electron. Write the half-reactions and balanced equation.

_____

_____

4. Write the half-reactions and a balanced equation for the reaction between tin(II) ions, Sn^{2+}, and permanganate ions, MnO_4^- (Mn has a charge of +7), in an acid solution to produce tin(IV) ions, Sn^{4+}, and manganese(II) ions, Mn^{2+}.

_____

_____

5. Steel screws can be plated with cadmium to minimize rusting. One reaction that can occur is written below:

$$Cd^{2+} + Ni \rightarrow Cd + Ni^{2+}$$

From this balanced equation, determine the half-reactions and write them out.

_____

_____

Unit 5

PUTTING IT ALL TOGETHER
A CHEMICAL PLANT FOR RIVERWOOD?

Riverwood Town Council Meeting Instructions and Responsibilities

Some members of your class will play the role of Riverwood Town Council members and will moderate the discussion of the question "A Chemical Plant for Riverwood?" Other students will be assigned to be members of the various groups outlined in your book. *Please note: this is a role-modeling activity! Get into your roles!* Use as many techniques as you can think of—role playing, costumes, banners, slogans, etc.—to present your group's position effectively.

Each group will have a specified time to present their case. Your group may choose several possible approaches:

1. **Analytical:** The analytical approach attempts to answer questions like what problems could be caused by the plant, what are the positive and negative aspects for either plant, and does the community really need a chemical plant in Riverwood?

2. **Defensive:** The defensive approach responds to criticisms that say one plant or the other should not be located in Riverwood. Your arguments would explain why the company will not be a problem to the city.

3. **Offensive:** The offensive approach supports locating a particular plant in Riverwood, using positive reasoning.

4. **Other:** Your group may prefer to use some combination of the approaches. Or your group may devise another argumentative style.

Each special interest group must first elect a "spokesperson." This person will address the town council and present the group's arguments to the other Riverwood citizens. Each group also must select an "examiner." This person will question the arguments of other groups.

Group members will then decide what position the group will take on the plant issue and come up with five arguments to support their position. The group must also prepare one question to ask each of the other groups. (Prepare alternative questions in case another group asks your question before your examiner has a chance to speak.) Use the worksheets that follow as your team prepares. During the presentations, every Riverwood citizen must fill out the forms below, stating the position of his or her group and the arguments and questions for the other groups.

After all groups have made their presentations and answered questions, there will be a one-minute period of time for rebuttal. A rebuttal is your group's chance to clarify your position, defend your actions, explain why your ideas are right, or accuse another group of falsifying the facts. The spokesperson will address the rebuttal—unless your group has selected a third person for this job. After all of the arguments and rebuttals, the town council members will adjourn to make a decision which plant (if any) will be located in Riverwood and to suggest a course for further action.

Finally, after the council renders its decision, each person is asked to do a wrap-up activity. As citizens of Riverwood, all students will write a letter to the editor of the *Riverwood News*, discussing the meeting and why you agree or disagree with the council's decision. The letter should be approximately 100 words long. As you write, remember that you are writing as a citizen of the city.

Name _____ Period _____ Date _____

NOTES ON THE DECISIONS OF MY GROUP
A CHEMICAL PLANT FOR RIVERWOOD?

My Special Interest Group: _____

Spokesperson: _____

Examiner: _____

Other group member(s):

_____

_____

Position on the plant issue:

_____

_____

_____

_____

Argument 1:

_____

_____

Argument 2:

_____

_____

Argument 3:

_____

_____

Argument 4:

_____

_____

Argument 5:

_____

_____

QUESTIONS MY TEAM WILL ASK OTHER GROUPS

A Chemical Plant for Riverwood?

Special Interest Group _____

Town Council

    1. _____

    2. _____

EKS Nitrogren Products Company

    1. _____

    2. _____

WYE Metals Corporation

    1. _____

    2. _____

Riverwood Industrial Development Authority

    1. _____

    2. _____

Riverwood Environmental League

    1. _____

    2. _____

Riverwood Taxpayer Association

    1. _____

    2. _____

Additional Groups (if necessary)

Title: _____

    1. _____

    2. _____

Title: _____

    1. _____

    2. _____

Title: _____

    1. _____

    2. _____

Unit 5

PUTTING IT ALL TOGETHER
RIVERWOOD TOWN COUNCIL MEETING EVALUATION

As you prepare for the group presentations, use the six topics listed below as guidelines. Your team's performance will be evaluated according to these standards. Then, as you listen to each team's presentation, use the guidelines to evaluate it as well. You may also evaluate your own group's performance according to these guidelines. Note that the grading system goes from 0 (poor performance) to 10 (excellent performance).

		Score
Group Arguments and Questions: From paper submitted by group, accuracy of facts, complete presentation of position.	Poor Great 0 4 8 12 16 20	
Presentation and Organization: Presentation was organized; students got into their roles and modeled the situation well; all students were actively involved.	Poor Great 0 4 8 12 16 20	
Quality of Rebuttals: Responses to statements given during the meeting.	Poor Great 0 4 8 12 16 20	
Creativity: Uniqueness of presentation.	Poor Great 0 4 8 12 16 20	
Visual Aids and Name Tags: Uniqueness of design, props, costumes, posters, transparencies, etc.	Poor Great 0 2 4 6 8 10	
Audience Participation: Involvement of group members in classroom presentations.	Poor Great 0 2 4 6 8 10	
Total (100 points possible)		

Unit 6

A.4 BUILDING SKILLS SUPPLEMENT

1. What experimental evidence led Rutherford to conclude each of the following?

   a. The nucleus of the atom contains most of the atomic mass.

   _____

   _____

   _____

   b. The nucleus of the atom is positively charged.

   _____

   _____

   c. The atom consists of mostly empty space.

   _____

   _____

2. In what ways are isotopes alike? In what ways are they different?

   _____

   _____

   _____

3. An atom of an element has an atomic number of 24 and a mass number of 52. What are the isotopic notation, the name, and the number of protons, neutrons, and electrons in the atom?

   _____

   _____

   _____

4. If an atom of an element has a mass number of 201 and has 121 neutrons, what is the isotopic notation of the element?

   _____

5. Give the isotopic notation for

   a. an atom containing 27 protons, 32 neutrons, and 25 electrons.   _____

   b. an atom containing 53 protons, 74 neutrons, and 54 electrons.   _____

6. Are the following statements true or false?

   a. _____ An element with an atomic number of 29 has 29 protons, 29 neutrons, and 29 electrons.

   b. _____ An atom of the isotope ^{60}Fe has 34 neutrons.

   c. _____ An atom of ^{31}P contains 15 protons, 16 neutrons, and 31 electrons.

   d. _____ ^{23}Na and ^{24}Na are isotopes.

   e. _____ ^{24}Na has one more electron than ^{23}Na.

   f. _____ ^{24}Na has one more neutron than ^{23}Na.

7. How many neutrons are in each of the following isotopes?

   a. titanium-46 _____

   b. plutonium-242 _____

   c. seaborgium-263 _____

   d. tungsten-186 _____

Unit 6

A.5 BUILDING SKILLS SUPPLEMENT: FINDING AVERAGE ATOMIC WEIGHT

Example: There are three isotopes of neon found on the earth: ^{20}Ne that accounts for 90.92% of the total, ^{21}Ne, which makes up 0.2571%, and ^{22}Ne, which makes up 8.822%. What is the average atomic weight of Ne?

Atomic weight = (weight of isotope) × (decimal of percent) + (weight of isotope) × (decimal of percent) + ...
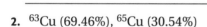
= (20) × (0.9092) + (21) × (0.002571) + (22) × (0.08822)
= 18.1840 + 0.053991 + 1.94084
= 20.1788 (rounded to least decimal place)

1. ^{10}B (19.78%), ^{11}B (80.22%)

_____

2. ^{63}Cu (69.46%), ^{65}Cu (30.54%)

_____

3. ^{69}Ga (60.27%), ^{71}Ga (39.73%)

_____

4. ^{79}Br (50.42%), ^{81}Br (49.58%)

_____

5. ^{85}Rb (72.10%), ^{87}Rb (27.90%)

_____

6. ^{35}Cl (77.30%), ^{37}Cl (22.70%)

_____

7. ^{107}Ag (51.72%), ^{109}Ag (48.28%)

_____

8. ^{28}Si (92.2%), ^{29}Si (4.7%), ^{30}Si (3.1%)

_____

9. ^{24}Mg (78.70%), ^{25}Mg (10.13%), ^{26}Mg (11.17%)

_____

10. ^{112}Sn (0.95%), ^{114}Sn (0.65%), ^{115}Sn (0.34%), ^{116}Sn (14.24%), ^{117}Sn (7.57%), ^{118}Sn (24.01%), ^{119}Sn (8.58%), ^{120}Sn (32.97%), ^{122}Sn (4.71%), ^{124}Sn (5.98%)

_____

_____

- # Unit 6

B.2 MAKING DECISIONS SUPPLEMENT: YOUR ANNUAL IONIZING-RADIATION DOSE

Common Sources of Radiation	Your Annual Dose (mrem)
1. Where You Live	
a. Cosmic Radiation (from outer space)	
Your exposure depends on elevation. These are annual doses.	
sea level 26 mrem 4000–5000 ft 47 mrem 0–1000 ft 28 mrem 5000–6000 ft 52 mrem 1000–2000 ft 31 mrem 6000–7000 ft 66 mrem 2000–3000 ft 35 mrem 7000–8000 ft 79 mrem 3000–4000 ft 41 mrem 8000–9000 ft 96 mrem	_____ mrem
b. Terrestrial Radiation (from the ground)	
If you live in a state bordering the Gulf or Atlantic coasts 16 mrem	
If you live in AZ, CO, NM, or UT 63 mrem	
If you live anywhere else in the continental United States 30 mrem	_____ mrem
c. House Construction	
If you live in a stone, adobe, brick, or concrete building 7 mrem	_____ mrem
d. Power Plants	
If you live within 50 miles of a nuclear power plant 0.009 mrem	
If you live within 50 miles of a coal-fired power plant 0.03 mrem	_____ mrem
2. Food, Water, Air	
Internal Radiation (based on average values)	
From food (carbon-14 and potassium-40) and 40 mrem from water (radon dissolved in water)	_____ mrem
From air (radon) 200 mrem	_____ mrem
3. How You Live	
Weapons test fallout* 1 mrem	_____ mrem
Travel by jet aircraft (per hour in air) 0.5 mrem	_____ mrem
If you have porcelain crowns or false teeth 0.07 mrem	_____ mrem
If you wear a luminous wristwatch 0.06 mrem	_____ mrem
If you go through airport security (each time) 0.002 mrem	_____ mrem
If you watch TV* 1 mrem	_____ mrem
If you use a video display (computer screen)* 1 mrem	_____ mrem
If you live in a dwelling with a smoke detector 0.008 mrem	_____ mrem
If you use a gas camping lantern with an old mantle 0.2 mrem	_____ mrem
If you wear a plutonium-powered pacemaker 100 mrem	_____ mrem
4. Medical Uses (radiation dose per procedure)	
X-rays: Extremity (arm, hand, foot, or leg) 1 mrem	_____ mrem
Dental 1 mrem	_____ mrem
Chest 6 mrem	_____ mrem
Pelvis/hip 65 mrem	_____ mrem
Skull/neck 20 mrem	_____ mrem
Barium enema 405 mrem	_____ mrem
Upper GI 245 mrem	_____ mrem
CT scan (head and body) 110 mrem	_____ mrem
Nuclear medicine (e.g., thyroid scan) 14 mrem	_____ mrem
Your Estimated Annual Radiation Dose	_____ mrem

*The value is less than 1 mrem, but adding that value would be reasonable.
Adapted from "Estimate your personal annual radiation dose," American Nuclear Society, 2000.

Questions

1a. _____

b. _____

2a. _____

_____

_____

b. _____

_____

3a. _____

_____

_____

b. _____

_____

_____

Unit 6

B.4 LABORATORY ACTIVITY: ALPHA, BETA, AND GAMMA RADIATION—PROCEDURE

Purpose

The purpose of this activity is to understand the penetrating ability, effect of distance, and shielding effects on alpha and beta particles, and gamma radiation.

Procedure

Part 1: Penetrating Ability

1. Prepare a data table with four data columns and three data rows for recording the number of counts detected per minute. The heading of each column should represent the type of shielding used, while each row should be labeled with the type of radiation being used.

2. Set up the apparatus shown in Figure 19 (page 435). There should be room between the source and the detector for several sheets of glass or metal.

3. Turn on the counter; allow it to warm up for at least 3 minutes. Determine the intensity of the background radiation by counting clicks for one minute in the absence of any specific radioactive sources. Identify and record this background radiation in counts per minute (cpm) below your data table.

4. Put on protective gloves. Using forceps, place a gamma radiation source on the ruler at a point where it produces a high reading on the meter (Figure 19). Observe the meter for 30 seconds and estimate the average cpm detected during this period. Record that cpm value. Then subtract the background reading from that value and record the corrected result.

5. Without moving the radiation source, place a piece of cardboard (an index card) between the detector and the source, as shown in Figure 20 (page 436).

6. Again observe the meter for 30 seconds. Record the average reading. Then correct that reading for background radiation and record the corrected result.

7. Repeat Steps 5 and 6 replacing the cardboard with a glass or plastic plate.

8. Repeat Steps 5 and 6 replacing the cardboard with a lead plate.

9. Repeat Steps 4 through 8 using a beta-particle source.

10. Repeat Steps 4 through 8 using an alpha-particle source.

Part 2: Effect of Distance on Intensity

11. Prepare a data table for recording your Part 2 observations.

12. Place a radioactive source designated by your teacher at the point on the ruler that produces nearly a full-scale reading (usually a distance of about 5 cm).

13. Measure the average reading over 30 seconds. Determine the average counts per minute and record that value. Then correct this reading by subtracting the background value. Record your corrected value (cpm) in the data table.

14. Move the source so its distance from the detector is doubled.

15. Again measure the average reading over 30 seconds. Record that initial value and the corrected value (cpm).

16. Move the source twice more, so the original distance is first tripled, then quadrupled, recording the initial and corrected reading after each move. (For example, if you started at 2 cm, you would take readings at 2, 4, 6, and 8 cm.)

17. Prepare a graph, plotting the corrected cpm on the *y* axis and the distance from source to detector (in cm) on the *x* axis.

Part 3: Shielding Effects

18. Prepare a data table containing two columns—one for glass and one for lead. The table should include three rows, corresponding to the number of sheets of material. Remember that you are starting with zero sheets.

19. Using forceps, place a source designated by your teacher on the ruler at a distance that produces nearly a full-scale reading.

20. Take an average reading over 30 seconds. Determine the counts per minute, correcting for background radiation. Record your corrected value on the table.

21. Place one glass plate between the source and the detector. Do not change the distance between the detector and the source. Take an average reading over 30 seconds. Then determine and record the corrected counts per minute.

22. Place a second glass plate between the source and the detector. Take an average reading over 30 seconds. Determine and record the corrected counts per minute.

23. Repeat Steps 21 and 22, using lead sheets rather than glass plates.

24. Wash your hands thoroughly before leaving the laboratory.

B.4 LABORATORY ACTIVITY: ALPHA, BETA, AND GAMMA RADIATION

Part 1: Penetrating Ability

Background radiation _____ cpm (counts per minute)

DATA TABLE	No Shielding	Cardboard	Glass/plastic plate	Lead plate
Gamma source				
Beta source				
Alpha source				

Note: Subtract background radiation before entering radiation data in the table.

Questions

1. _____

2. _____

_____

_____

_____

3a. _____

b. _____

4. _____

_____

Part 2: The Effect of Distance on Intensity

DATA TABLE				
Distance				
Corrected cpm				

Note: Subtract background radiation before entering radiation data in the table.

Construct your graph in the space on p. 174.

Questions

5. _____

6. _____

7. _____

_____

Part 3: Shielding Effects

DATA TABLE		
Sheets of material	**Glass**	**Lead**
Zero sheets		
One thickness		
Two thickness		

Note: Subtract background radiation before entering radiation data in the table.

Questions

8a. _____

b. _____

9a. _____

b. _____

_____

10a. _____

b. _____

_____

_____

Unit 6

B.5 BUILDING SKILLS SUPPLEMENT: NUCLEAR BALANCING ACT

Write a balanced equation for each of the following nuclear reactions.

Examples:

Chlorine-36 decays by beta emission

$$^{36}_{17}Cl \longrightarrow ^{36}_{18}Ar + ^{0}_{-1}e$$

Dubnium-262 decays by alpha emission

$$^{262}_{105}Db \longrightarrow ^{258}_{103}Lr + ^{4}_{2}He$$

1. Krypton-87 decays by beta emission.

2. Curium-240 decays by alpha emission.

3. Uranium-232 decays by alpha emission.

4. Silicon-32 decays by beta emission.

5. Zinc-71 decays by beta emission.

6. Americum-243 decays by alpha emission.

7. Cobalt-60 decays by beta emission.

8. Phosphorous-32 decays by beta emission.

9. Gadolinium-150 decays by alpha emission.

10. Lead-210 decays by emitting both a beta and an alpha particle.

Unit 6

B.8 LABORATORY ACTIVITY: CLOUD CHAMBERS—PROCEDURE

The cloud chamber you will use consists of a small plastic container and a felt band moistened with 2-propanol (isopropyl alcohol). The alcohol evaporates faster than water and saturates air more readily. The cloud chamber will be chilled with dry ice to promote supersaturation and cloud formation.

Procedure

1. Fully moisten the felt band inside the cloud chamber with alcohol. Also place a small quantity of alcohol on the container bottom.

2. Using gloves and forceps, quickly place the radioactive source in the chamber. Replace the lid.

3. To cool the chamber, place it on a flat surface of crushed dry ice. Ensure that the chamber remains level.

4. Leave the chamber on the dry ice for three to five minutes.

5. Dim or turn off the room lights. Focus the light source at an oblique angle (not straight down) through the container so that the chamber base is illuminated. If you do not observe any vapor trails, try shining the light through the side of the container.

6. Observe the air in the chamber near the radioactive source. Record your observations.

B.8 LABORATORY ACTIVITY: CLOUD CHAMBERS

Purpose

The purpose of this activity is to observe particle trails from a radioactive source.

Observations inside the cloud chamber

_____

_____

_____

_____

_____

_____

_____

_____

Observations near the radioactive source

_____

_____

_____

Questions

1. _____

_____

2. _____

_____

_____

3. _____

_____

_____

Unit 6

C.1 BUILDING SKILLS SUPPLEMENT: HALF LIFE

1. The half-life of sulfur-38 is 2.87 hours.

   a. After 8.61 hours, what percent of the original radiation is left?

   _____

   b. If 3.125% of the original radiation is being emitted, how many half-lives have passed?

   _____

2. Calcium-47 has a half-life of 4.5 days. Graph the decay of a sample for 7 half-lives, with time (days) on the x axis and amount of original radiation (%) on the y axis.

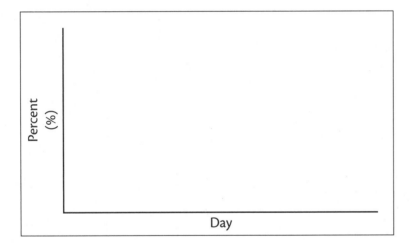

   a. How much calcium-47 radiation remains after two weeks?

   _____

   b. When will the calcium radiation be 10% of the original value?

   _____

   c. How much of the sample will have decayed after 7 half-lives?

   _____

3. If 1250 counts of a 10 000 count radioactive sample is being emitted after one day, what is the half-life of the element?

   _____

4. Strontium-90 has a half-life of 28.8 years. How long will it take for a 5-count sample to decay to 0.3125 counts?

_____

5. Phosphorous-32 has a half-life of 14.3 days. A 30-count sample is stored for 114.4 days. How much radiation of the original sample remains?

_____

6. The half-life of carbon-14 is 5730 years. If a sample of wood containing 60 counts of carbon-14 when alive now only emits 3.75 counts, how old is the sample?

_____

7. Germanium-66 has a half-life of 2.5 hours. After 10 hours, if only 25 counts of an original sample are emitted, what was the radiation of the original sample?

_____

Unit 6

C.4 BUILDING SKILLS SUPPLEMENT

Write a balanced equation for each of the following nuclear reactions.

Example: Oxygen-16 plus a neutron results in the formation of another element and the release of an alpha particle.

$$^{16}_{8}O + ^{1}_{0}n \rightarrow ^{13}_{6}C + ^{4}_{2}He$$

1. Boron-10 plus a neutron results in the formation of another element and the release of an alpha particle.

   _____

2. Beryllium-9 plus a proton results in the formation of another element and the release of an alpha particle.

   _____

3. Einsteinium-253 plus an alpha particle results in the formation of another element and the release of a neutron.

   _____

4. Lithium-7 plus a proton results in the formation of another element and the release of a neutron.

   _____

5. Plutonium-241 plus another particle results in the formation of plutonium-242 and the release of gamma rays.

   _____

6. Argon-40 plus an alpha particle produces another element and the release of a neutron.

   _____

7. Einsteinium-252 was bombarded by a berylium-9 atom, producing a new element and three neutrons.

   _____

8. Plutonium-239 can be produced by bombarding uranium-238 with an alpha particle. Some neutrons are released.

   _____

9. Uranium-235 is bombarded with a neutron to produce tellurium-137, another element, and two neutrons.

_____

10. On the sun, three steps are needed to create helium from hydrogen (nuclear fusion). In the third step, two helium-3 atoms react to form helium-4 and two hydrogen atoms. (Since the atomic mass of the helium-4 is less than that of the four hydrogen atoms used to make it, the difference in mass accounts for the energy released from the sun.)

_____

_____

● Unit 6

PUTTING IT ALL TOGETHER SUPPLEMENT
THE TRUTH ABOUT NUCLEAR CHEMISTRY

As you prepare for the group presentations, use the seven topics listed below as guidelines, which will form the basis of your team's evaluation. Note the emphasis on the scientific content of your presentation. You may use the guidelines to evaluate the performance of other teams as well as your own.

COMMUNICATING SCIENTIFIC AND TECHNOLOGICAL KNOWLEDGE		
Specification	Comments	Score
Presentation was clear and understandable; presenter(s) demonstrated a professional posture and maintained visual contact with the audience.		(20)
Determinations of the statements made in the flyer were stated as true, false, or partially true.		(20)
An explanation was provided for each determination, based on applicable science or technology.		(20)
Replacement statements to the flyer accurately addressed the issue, were well written and produced. (Superior work would include a replacement flyer with new statements.)		(10)
Visual aids, such as graphs, power point presentations, transparencies, charts, or graphics, were used well.		(10)
Everyday examples and applications were used to clarify the issues to the audience. (Superior presentations made the everyday examples appropriate for the senior-citizen audience.)		(10)
Presenter(s) responded to questions about the presentation in an informed and professional manner.		(10)
Total (100 points possible)		

Unit 7

A.2 MAKING DECISIONS: DIET AND THE FOOD PYRAMID

Part 1

To make an accurate food diary, you must list all food items, snacks, beverages (even water), and dietary supplements that you eat for the next three days. Estimate the amount of each item eaten by number or units, total mass, or total volume. Whenever possible, use food labels to help you estimate the number of servings. The following discussion of serving size will help you fill in your diary accurately.

Food Guide Pyramid Serving Size Guide

For each food group in the Food Guide Pyramid, the unit of measure is the serving. Learning to judge serving sizes takes a little practice because each item in your meal may add up to *more* than one serving. For example, a bowl of cereal may contain 2 or 3 servings. A 20-ounce bottle of soda pop contains 2 1/2 servings.

That means that you must be careful—serving size does not equal your portion size! The following tables illustrate one way that a person could consume the daily requirements of each food group during one day.

Food Group	Examples of Food Required for One Serving
Breads, Cereals, Grains, and Pasta Group (6–11 servings)	1 slice of bread or 1 dinner roll 1 ounce of ready-to-eat cereal 1/2 cup of cooked cereal, rice, or pasta
Vegetables Group (3–5 servings)	1 cup of raw leafy vegetables 1/2 cup of other vegetables (cooked or raw) 3/4 cup of vegetable juice
Fruits Group (2–4 servings)	1 medium apple, banana, or orange 1/2 cup of chopped, cooked, or canned fruit 1/4 cup of dried fruit 3/4 cup of 100% fruit juice (not fruit drink)
Meat, Poultry, Fish, Dry Beans, Eggs, and Nuts Group (2–3 servings)	2 to 3 ounces cooked lean meat, cooked poultry, cooked fish 1/2 cup of cooked dry beans 1 egg 4 to 6 tablespoons of peanut butter
Milk, Yogurt, and Cheese Group (2–3 servings)	1 cup of milk or yogurt 1-1/2 ounces of natural cheese 2 ounces of processed cheese

Calculating Serving Sizes

When in doubt, you can quickly estimate the size of a serving by using these visual examples.

- *3 ounces of meat, poultry, or fish:* a deck of playing cards or a cassette tape
- *1 cup of fruit or yogurt:* a baseball
- *1/2 cup of chopped vegetables:* three regular ice cubes
- *1 medium potato:* a computer mouse
- *1 cup of potatoes, rice, or pasta:* the size of a fist or a tennis ball
- *1 cup chopped fresh leafy greens:* 4 lettuce leaves
- *2 Tablespoons of peanut butter:* a golf ball
- *1 ounce of cheese:* four dice or a tube of lipstick
- *1/2 cup of cooked vegetables:* 7–8 baby carrots or carrot sticks, 1 ear of corn

For identifying the Food Guide Pyramid Groups on the Food Inventory, use this guide:

1. **Bread, cereal, grains, and pasta**
2. **Vegetables**
3. **Fruits**
4. **Milk, yogurt, and cheese**
5. **Meat, poultry, fish, dry beans, eggs, and nuts**
6. **Fats, oils, and sweets**

Name _____ Period _____ Date _____

Day # _____	Quantity	Number of Servings	Food Guide Pyramid Group(s)
Breakfast			
Lunch			
Dinner			
Snacks and Dietary Supplements			

Unit 7

A.2 MAKING DECISIONS: DIET AND THE FOOD PYRAMID

Part 2

Use the data collected from the three-day food inventory to fill out the following table.

DATA TABLE

Food Guide Pyramid Group	Day 1 Servings	Day 2 Servings	Day 3 Servings	Total Servings for 3 Days	Average Number of Daily Servings
Bread, cereal, grains, and pasta (1)					
Vegetables (2)					
Fruits (3)					
Milk, yogurt, and cheese (4)					
Meat, poultry, fish, dry beans, eggs, and nuts (5)					
Fats, oils, and sweets (6)					

Questions

3a. _____

_____

b. _____

_____

c. _____

_____

4. _____

_____

_____

5. Item 1 _____

   a. Group assigned _____

   b. _____

     _____

   c. _____

     _____

Item 2 _____

   a. Group assigned _____

     _____

   b. _____

     _____

   c. _____

     _____

Item 3 _____

   a. Group assigned _____

     _____

   b. _____

     _____

   c. _____

     _____

6. _____

  _____

  _____

  _____

7. _____

  _____

  _____

Unit 7

A.3 LABORATORY ACTIVITY:
ENERGY CONTAINED IN A SNACK—PROCEDURE

In this laboratory activity you will determine the energy contained in a snack-food item. Use the candle-burning procedure (Unit 2, pages 203–205) and information in the Calculations section of your text to guide you as you design a procedure for measuring the energy contained in a snack.

A.3 LABORATORY ACTIVITY:
ENERGY CONTAINED IN A SNACK

Purpose

The purpose of this activity is to determine the energy contained in a snack-food item.

Procedure

_____

_____

_____

_____

_____

_____

_____

_____

_____

_____

_____

_____

_____

Unit 7

B.5 MAKING DECISIONS: ANALYZING FATS AND CARBOHYDRATES

Carbohydrates and fats are the main sources of energy in the foods you eat. Using the food inventory you created for Activity A.2, determine the mass of carbohydrates and fats contained on the inventory.

1. Using appropriate resources, determine the mass of carbohydrates (in grams) contained in each item on the three-day food inventory. Then, calculate the percent of Calories provided by carbohydrates. Record both sets of data.

2. Using appropriate resources, determine the mass of fat (in grams) contained in each item on the three-day food inventory. If possible, identify the total mass of saturated and unsaturated fat in each item. Record these data.

3. Calculate the average mass of fat supplied each day.

4. a. Calculate the average daily energy (Calories) supplied by the fat.

   b. Based on that value, is food energy from fat less than 30% of the total food energy supplied?

5. Identify possible ways to reduce the quantity of fat in the food inventory you are analyzing.

FOOD INVENTORY

Day # _____	Carbo-hydrates (grams)	Total Fat (grams)	Saturated Fat (grams)	Unsaturated Fat (grams)
Breakfast				

FOOD INVENTORY

Day # _____	Carbo-hydrates (grams)	Total Fat (grams)	Saturated Fat (grams)	Unsaturated Fat (grams)
Lunch				
Dinner				
Snacks and Dietary Supplements				
Total for the Day				

Summary

1. *Carbohydrates*

Day 1—Mass of Carbohydrates	
Day 2—Mass of Carbohydrates	
Day 3—Mass of Carbohydrates	
Total Mass of Carbohydrates in the Inventory	
Total Calories in Inventory (Section A.2)	
% of Calories Provided by Carbohydrates	

2 & 3. *Fats*

Day 1—Mass of Fats	
Day 2—Mass of Fats	
Day 3—Mass of Fats	
Total Mass of Fats in the Inventory	
Average Mass of Fats Each Day	

4a. Average daily energy supplied by fat _____

b. Average daily total energy (in Calories) from A.2 inventory _____

Percentage of total energy supplied by the fat in the inventory _____

5. _____

_____

_____

Unit 7

C.3 LABORATORY ACTIVITY: ENZYMES—PROCEDURE

You will be assigned a particular material—apple, potato, or liver—to examine for the presence of the enzyme catalase. You will test a fresh piece of the material and one that has been boiled, to see whether either material catalyzes the decomposition of hydrogen peroxide.

Read the procedure. Prepare a data table with appropriately labeled columns for your data.

Procedure

1. Obtain two pieces of your assigned food sample—one piece that is fresh and one that has been boiled.

2. Label two 16 mm × 125 mm test tubes—one "fresh," the other "boiled."

3. Add 5 mL 3% hydrogen peroxide, H_2O_2, solution to each test tube. Observe whether any bubbles form—that is, whether a gas is produced. (Why is that a sign of hydrogen peroxide decomposition?)

4. Add a portion of the fresh material to the appropriate test tube. Insert a stopper containing a segment of glass tubing into the mouth of the test tube, and arrange the tubing as shown in Figure 28. Be sure the end of the glass tubing is submerged in the beaker of water.

5. For three minutes, record the estimated number of bubbles formed in the test tube each minute.

6. Add a portion of boiled material to the second test tube, and repeat Steps 4 and 5.

7. Discard and dispose of the materials and solutions as directed by your teacher.

8. Wash your hands thoroughly before you leave the laboratory.

C.3 LABORATORY ACTIVITY: ENZYMES

Purpose

The purpose of this activity is to investigate the rate of an enzyme-catalyzed reaction.

Assigned Food Sample _____

DATA TABLE

Bubbles formed	Fresh Test Tube	Boiled Test Tube
End of the First Minute		
End of the Second Minute		
End of the Third Minute		

Questions

1. _____

   _____

2. _____

   _____

3a. _____

   _____

b. _____

   _____

4. _____

   _____

5. _____

   _____

   _____

Unit 7

C.5 LABORATORY ACTIVITY: AMYLASE TESTS—PROCEDURE

Introduction

In this activity, you will consider how temperature and pH affect the performance of the enzyme amylase. Amylase in saliva breaks down starch molecules into individual glucose units. Glucose reacts with Benedict's reagent to produce a yellow-to-orange precipitate. The color and amount of precipitate formed is a direct indication of the concentration of glucose generated by the catalyst.

Procedure

Day 1: Preparing the Samples

1. Label each of five test tubes (near the top) with the temperature your group is assigned to investigate, and with pH values of 2, 4, 7, 8, and 10. Also mark each label so that on Day 2 your set of test tubes can be distinguished from those of other groups.

2. Using solutions provided by your teacher, add 5 mL of pH 2, 4, 7, 8, or 10 solutions to the appropriate tubes.

3. Add 2.5 mL of starch suspension to each tube.

4. Add 2.5 mL of 0.5% amylase solution to each tube.

5. Insert a stopper into each tube. Hold the stopper in place with your thumb or finger, and shake the tube well for several seconds.

6. Leave the tubes that are to remain at room temperature in the laboratory overnight as directed by your teacher.

7. Give your teacher the tubes to be refrigerated.

8. Wash your hands thoroughly before leaving the laboratory.

Day 2: Evaluating the Results

9. Prepare a hot-water bath by adding about 100 mL of tap water to a 250-mL beaker. Add a boiling chip. Heat the beaker on a hotplate.

10. Place 5 mL of Benedict's reagent in each tube. Replace each stopper, being careful not to mix them up. Securing the stopper in place with a thumb or finger, shake each tube well for several seconds.

11. Ensure that the tubes are clearly labeled. Remove the stoppers, and place the tubes into the water bath.

12. Heat the test tubes in the hot-water bath until the solution in at least one tube has turned yellow or orange. Then continue heating for 2–3 minutes more.

13. Use a test tube holder to remove the tubes from the water bath. Arrange them in order of increasing pH.

14. Observe and record the color of the contents of each tube.

15. Share your data with your classmates as directed by your teacher.

16. Wash your hands thoroughly before leaving the laboratory.

C.5 LABORATORY ACTIVITY: AMYLASE TESTS

Purpose

The purpose of this activity is to explore how temperature and pH affect the performance of an enzyme.

DATA TABLE

Circle one:	Room Temperature	Refrigerated
pH	Observation of Test Tube	
2		
4		
7		
8		
10		

Which test tube first turned yellow or orange? _____

Questions

1a. _____

b. _____

2. _____

_____

_____

● # Unit 7

C.6 MAKING DECISIONS: PROTEIN CONTENT

You have analyzed your three-day food inventory in terms of energy it provides and the fat and carbohydrate molecules it delivers. Now consider whether that food inventory meets recommended amounts of a key building block of living material.

Proteins are the key building blocks of living material. Using your A.2 inventory, find the total mass of proteins in the foods you eat.

Food Inventory Day # _____	Proteins (grams)	Food Inventory Day # _____	Proteins (grams)	Food Inventory Day # _____	Proteins (grams)
Breakfast					
Lunch					
Dinner					
Snacks and Dietary Supplements					
Total for the Day					

1. Total mass of protein for three days: _____

Average mass of protein per day: _____

2. _____

_____

_____

Unit 7

D.2 LABORATORY ACTIVITY: VITAMIN C—PROCEDURE

This laboratory procedure is based on chemical properties of ascorbic acid (vitamin C) and iodine. You will perform a titration, a common procedure used to determine the concentrations of substances in solutions. A known amount of one reactant will be added slowly from a Beral pipet to another reactant in a wellplate until just enough has been added for a complete reaction. Completion of the reaction—the endpoint—is noted by a color change or other highly visible change. Knowing the chemical equation for the reaction involved, you can then calculate the unknown amount of the second reactant from the known amount of the first reactant.

Procedure

Part 1: Finding the Iodine Solution Concentration

1. Fill a Beral pipet with vitamin C solution provided by your teacher. Then determine how many drops of vitamin C solution delivered by that pipet represent a volume of 1.0 mL.

2. Fill a second Beral pipet with iodine solution. Determine the number of drops from that pipet that represent a 1.0-mL volume, just as you did in Step 1.

3. Add 25 drops vitamin C solution to a well of a clean 24-well wellplate. The vitamin C concentration is 1.0 mg vitamin C/mL.

4. Add one drop of starch solution to the well.

5. Place a piece of white paper underneath the wellplate—this will help you detect the color. Add iodine solution one drop at a time to the well containing the starch/vitamin C mixture. After each addition of iodine solution, gently stir the resulting mixture with a toothpick. Count the drops of iodine solution added to reach the endpoint (the first "permanent" blue-black color). Continue adding iodine solution drop by drop while stirring, until the solution in the well remains blue-black for 20 seconds. If the color fades before 20 seconds have elapsed, add another drop of iodine solution. Record the number of drops of iodine needed to reach the endpoint.

6. You know that the concentration of the vitamin C solution used is 1.0 mg/mL. How many milligrams of vitamin C react with one drop of iodine solution? In this procedure, 25 drops of vitamin C solution reacted with the iodine solution. How many drops of iodine solution did it take to react with that much vitamin C solution? How do drops of vitamin C solution relate to milligrams of vitamin C? Use that information to calculate how many milligrams of vitamin C react with one drop of iodine solution.

Part 2: Analyzing Beverages for Vitamin C

7. Using a Beral pipet, add 25 drops of a beverage to a clean well of a wellplate.

8. Add one drop of starch solution to the well.

9. With a Beral pipet, add iodine solution one drop at a time to the well containing the beverage/starch mixture. After each addition, stir the resulting mixture with a toothpick. Count the total drops of iodine solution that are added to reach the endpoint (a "permanent" blue-black color). (Note: Colored beverages may not produce a true blue-black endpoint color. For example, red beverages may appear purple at the endpoint.)

10. Continue adding iodine solution drop by drop, while stirring, until the solution in the well remains blue-black for 20 seconds. If the color fades before 20 seconds have elapsed, add another drop of iodine solution. Record the total drops of iodine solution needed to reach the endpoint.

11. Repeat Steps 7 through 10 for each beverage you are assigned to analyze.

12. Use data from Step 10 and the results of the calculation in Step 6 to calculate the milligrams of vitamin C contained in 25 drops of each assigned beverage.

13. Rank the beverages from the highest level to the lowest level of vitamin C.

14. Wash your hands thoroughly with soap and water before leaving the laboratory.

D.2 LABORATORY ACTIVITY: VITAMIN C

Purpose

The purpose of this activity is to determine the amount of vitamin C in common beverages and rank them according to their vitamin C concentration.

Note: Step 6 Calculation:
Follow the following equation to determine the mg of vitamin C in the various beverages.

$$25 \text{ drops} \times \frac{1 \text{ mL Vitamin C}}{\underline{\hspace{1cm}}\text{drops}} \times \frac{1 \text{ mg Vitamin C}}{1 \text{ mL Vitamin C}} \times \frac{1}{\underline{\hspace{1cm}}\text{drops } I_2} = \underline{\hspace{1cm}} \text{ mg Vitamin C/drop } I_2$$
$$\hspace{3.5cm} \text{(step 1)} \hspace{4cm} \text{(step 5)}$$

DATA TABLE

Beverage	Drops of Iodine Solution Used ×	Conversion Factor (step 6 =)	mg Vitamin C per 25 drops of beverage	Rank

Questions

1a. _____

_____

b. _____

_____

2. _____

_____

_____

_____

Unit 7

D.5 LABORATORY ACTIVITY: FOOD COLORING ANALYSIS—PROCEDURE

In this laboratory activity, you will analyze the food dyes in two commercial candies and compare them with the dyes in food coloring. You will separate and identify the food dyes with the help of paper chromatography. To analyze the results of this experiment, you will calculate the R_f value for each spot.

Before beginning the activity, prepare a data table that contains this information: The table should have six vertical columns labeled "Sample," "Initial color of sample," "Dyes observed," "Distance to dye spot (cm)," "Distance solvent moved (cm)," and "R_f value." The table should have three rows—one for each candy sample and one for the food coloring sample.

Procedure

1. Obtain one piece each of two different commercial candies from your teacher, as well as a food-coloring sample. The food coloring and candies should all be the same color.

2. Put each candy into a separate well of a wellplate. Note which candy is in each well. Add 5 to10 drops of water to each well. Stir the mixture in each well with a separate toothpick until the color completely dissolves from the candy. Add 3 to 4 drops of the food coloring sample to a third well in the wellplate. Observe and record the initial color of each sample.

3. Obtain a strip of chromatography paper, handling it only by its edges. With a pencil (not pen), draw one horizontal line 2 cm from the bottom of the paper and another horizontal line 3 cm from the top. Label the strip as shown in Figure 40.

4. Next, you will place a spot of each dye on the bottom line—the spots should not be large. To do this, use a separate toothpick for each of the three samples. Place a drop of the first candy's color solution with a toothpick as indicated in Figure 40. Allow the drop of solution to sit until the spot it makes stops spreading out on the paper. Then apply a second drop of the same sample on top of the spot. Using the same two-drop technique, apply drops of the second candy's color solution and the food-coloring sample to appropriate places on the pencil line, as indicated in Figure 40.

5. Obtain a chromatography chamber. Make a mark on the outside of the chamber that is approximately 1 cm from the bottom. Pour solvent into the chromatography vessel to the 1-cm mark.

6. Lower the spotted chromatography paper into the chromatography vessel until the bottom of the paper rests evenly in the solvent (water). Be sure the spots remain above the solvent surface. Place a lid on top of the chromatography chamber.

7. Allow the solvent to rise past the spots and up the paper until the solvent is close to the penciled line, about 3 cm from the top of the paper. Then remove the paper from the vessel and, using a pencil, mark the farthest point of travel by the solvent. Allow the paper to air-dry overnight.

8. Record the colors observed for the dye sample and candy solutions.

9. Measure the distance (in cm) from the initial pencil line where you placed the spots to the center of each dye spot. Record these distances in your data table.

10. Measure the total distance (in cm) that the solvent moved.

11. Calculate the R_f value of each dye spot you observed in your samples. Record these values.

D.5 LABORATORY ACTIVITY: FOOD COLORING ANALYSIS

Purpose

The purpose of this activity is to separate and identify food dyes in food colors using paper chromatography.

DATA TABLE

Sample	Color of initial sample	Colors observed	Distance from pencil line to center of dye spot (cm)	Total distance the solvent moved (cm)	R_f value
Food coloring					

Name _____   Period _____   Date _____

Questions

1. _____
_____

2. _____

3. _____

4. _____

5a. _____
_____

b. _____
_____

6a. _____
_____

b. _____
_____

c. _____
_____
_____
_____

Unit 7

D.9 MAKING DECISIONS SUPPLEMENT: ANALYZING VITAMINS AND MINERALS

In the last analysis of your food inventory, you will take a look at vitamins and minerals, essential in growth, reproduction, and good health. Use your A.2 food inventory to collect data on these essential ingredients of your health.

DATA TABLE

FOOD INVENTORY Day # _____	VITAMINS		MINERALS	
Identification of vitamins and minerals→				
Breakfast				
Lunch				

DATA TABLE

FOOD INVENTORY Day # _____	VITAMINS		MINERALS	
Dinner				
Snacks and Dietary Supplements				
Total for the Day				

Questions

3.

	VITAMINS		MINERAL	
Day 1 Total				
Day 2 Total				
Day 3 Total				
Total for 3 days				
Average intake				

4. _____

_____

_____

5. _____

_____

_____

_____

_____

_____

Unit 7

PUTTING IT ALL TOGETHER
FOOD DIARY PROJECT

INTRODUCTION

"You are what you eat." Is this saying true or false? This project will attempt to answer that question by taking a scientific look at the content of the foods you consume. The purpose of the PIAT is for you to evaluate the chemicals in your diet for quality, quantity, and effect of combination and energy content. You have already accomplished many of the objectives as you worked through this unit. Now is the time to bring all of your work together.

Objectives

- **Measure** the amounts of foods consumed over a period of time.
- **Compare** the caloric intake to a recommended amount.
- **Research** the chemical content of consumed foods.
- **Research** the recommended daily allowance of vitamins, minerals, and chemicals.
- **Research** the recommended amounts of energy required and expended by humans.
- **Compute** the average amount of food mass, energy, protein, carbohydrate, fat, vitamins, and minerals consumed per day.
- **Compare** quantities consumed to recommended quantities.

Materials

- Diary or logbook
- Access to research materials: library, Internet, reference books
- Typewriter, word processor, or computer, or as directed by your teacher

Procedure

1. Keep a detailed diary/logbook over a three-day period. Record the date, time, and quantity of measure for each type of food consumed.
2. Record your initial and final body mass in grams over the period of the diary.
3. Research the energy and chemical content of each of your foods.
4. Research the recommended daily allowance (RDA) for each food. Take into account any personal deviations due to health or medical conditions.

5. Research your recommended daily energy requirements.

6. Research methods for computing energy expenditures and compute your personal value.

7. Research the structure, sources, and function of the vitamins and minerals in your diet. Choose one of each to report on.

Results

1. Sum the total quantities of mass, energy, protein, carbohydrate, fat, vitamins, and minerals consumed over the test period.

2. Compute your average daily body-mass gain/loss, energy balance, protein quantity, carbohydrate quantity, fat quantity, vitamin quantity, and mineral quantity over the test period.

3. Make a data table that shows the recommended daily requirements next to your average values. (As an option, you can add bar graphs to display comparisons visually.) See pages 219–220 for a model of this table.

Analysis and Conclusions

1. Compare each computed average daily measurement to the recommended value.

2. Identify areas of satisfactory diet and areas where improvement is needed.

Summary

1. Evaluate how and where your personal diet and energy profile can be improved.

2. Make comments on the value of this project and how it might be improved.

Evaluation

The project will be evaluated for completeness in meeting all objectives and requirements of this assignment. A bibliography is required—use the format designated by your teacher.

FOOD INVENTORY

Day # _____	Food Energy (Calories) Making Decisions A.7	Carbohydrates (grams) Making Decisions B.5	Fat (grams) Making Decisions B.5	Protein (grams) Making Decisions C.6	Vitamins (milligrams) Making Decisions D.9	Minerals (milligrams) Making Decisions D.9
Breakfast						
Lunch						

(continued)

FOOD INVENTORY

Day # _____

	Food Energy (Calories) Making Decisions A.7	Carbohydrates (grams) Making Decisions B.5	Fat (grams) Making Decisions B.5	Protein (grams) Making Decisions C.6	Vitamins (milligrams) Making Decisions D.9	Minerals (milligrams) Making Decisions D.9
Dinner						
Snacks						
Total for the Day						

Unit 7

PUTTING IT ALL TOGETHER SUPPLEMENT
WHAT KIND OF FOOD AM I?

Suppose that while you are keeping your food diary one evening you decide to pop a quick dinner in the microwave. In the freezer you find two chicken dinners: "Chicken Supreme" and "Chicken Delight." Which one should you choose? Perhaps the nutritional information will help you decide.

Chicken Delight
Nutritional Facts
1 serving
Calories: 370
Calories from fat: 70
Percent Daily Values*
Total fat: 8 g (12%)
Saturated fat: 3 g (15%)
Cholesterol: 45 mg (15%)
Sodium: 470 mg (20%)
Total Carbohydrates: 53 g (18%)
Dietary fiber: 6 g (23%)
Sugars: 35 g
Protein: 23 g
Vitamin A: 30%
Vitamin C: 20%
Calcium: 10%
Iron: 15%

Chicken Supreme
Nutritional Facts
1 serving
Calories: 290
Calories from fat: 130
Percent Daily Values*
Total fat: 15 g (22%)
Saturated fat: 4 g (17%)
Cholesterol: 50 mg (17%)
Sodium: 900 mg (38%)
Total Carbohydrates: 27 g (9%)
Dietary fiber: 3 g (12%)
Sugars: 3 g
Protein: 14 g
Vitamin A: 6%
Vitamin C: 100%
Calcium: 67%
Iron: 10%

*Percent Daily Values are based on a 2000-Calorie diet. Your Daily Values may be higher or lower depending on your Calorie needs.

*Percent Daily Values are based on a 2000-Calorie diet. Your Daily Values may be higher or lower depending on your Calorie needs.

1. You decide to eat the "Chicken Delight."

   a. In order to avoid exceeding your daily maximum allowance of total fat, how many grams of fat could you have eaten earlier in the day prior to eating this meal?

   _____

   b. Suppose it turns out that you consumed 45 g of fat in the meals and snacks before you ate dinner. What percent of a 2000-Calorie Daily Value for total fat did you eat today, including dinner?

   _____

2. A friend comes over, checks the two labels, and tells you that you should have eaten the "Chicken Supreme."

   a. On a nutritional basis, justify to your friend your eating the "Chicken Delight."

   _____

   _____

   b. On what nutritional basis can your friend claim to be right?

   _____

   _____

3. If you need 15 mg of iron per day, how much iron did your dinner provide? _____

4. Female teenagers need 44 g of protein per day. Male teenagers need 59 g of protein per day. What percent did this dinner provide of your daily need?

   _____

5. What percentage of Calories comes from fat in a serving of "Chicken Delight"? _____

6. What percentage of Calories comes from fat in a serving of "Chicken Supreme"? _____

7. Advertisements for vitamin supplements frequently propose that their product will "boost your energy." Make a case in favor of or opposed to this claim.

   _____

   _____

   _____

8. Another friend claims to eat only natural foods because they contain no chemicals. What arguments could you present to show that your friend is wrong?

   _____

   _____

   _____

   _____

● # Unit 7

PUTTING IT ALL TOGETHER SUPPLEMENT
....
USING FOOD LABEL INFORMATION

One source of information for your food diary is the manufacturer's labels on packages of all kinds. Labels provide information about the ingredients used to make the product and nutritional information. Using the information on this fictitious breakfast cereal label, answer the following questions. Show the setup for any math question.

Ingredients: Whole oat flour, sugar, corn syrup, partially hydrogenated soybean oil, wheat starch, salt, cinnamon, trisodium phosphate, calcium carbonate, color and freshness preserved by sodium sulfite, sulfur dioxide, vitamin E, and BHT.

NUTRITION FACTS				
Serving size 1/2 cup (30g)				
Calories 120	Calories from fat 25			
	% Daily Values*			**% Daily Values***
Total fat 2.5 g	4%	Vitamin A		15%
Saturated fat 0 g	0%	Vitamin C		25%
Cholesterol 0 mg	15%	Calcium		2%
Sodium 190 mg	8%	Iron		25%
Potassium 35 mg	1%	Vitamin D		10%
Total Carbohydrates 23 g	8%	Thiamin		25%
Dietary fiber 2 g	12%	Riboflavin		25%
Sugars 13 g		Phosphorus		6%
Other Carbohydrate 8 g		Magnesium		4%
Protein 2 g		Niacin		25%
		Vitamin B$_6$		25%
		Folic Acid		25%
		Zinc		2%
		Copper		2%

*Percent Daily Values are based on a 2000-Calorie diet. Your Daily Values may be higher or lower depending on your Calorie needs.

1. What percent of the total Calories are from fat? _____

2a. Since there is no saturated fat, what types of fats are present in this cereal? _____

b. Which ingredient is the main source of fat in the cereal? _____

3. How many of the total Calories come from carbohydrates? _____

4. What percent of the total Calories are proteins? _____

5. How many mg of Vitamin C does this cereal provide? _____

6. How many mg of phosphorus are provided by this cereal? _____

7. Why do you think that the percentage of phosphorus is so low? _____

_____

8a. Which water soluble vitamins are listed? _____

_____

b. Which fat soluble vitamins are listed? _____

_____

9a. Which macrominerals are listed? _____

b. Which trace minerals are listed? _____

_____

10. What is the most abundant ingredient in this cereal? _____

_____

11. What is the least abundant ingredient in this cereal? What is it used for in the cereal? _____

_____

12. Name the additive that is used to preserve color and freshness. Write its formula. _____

_____

13. If you add the Calories for fats, carbohydrates, and proteins the total is greater than 120 Cal. Why do you think this is so?

_____

_____

Unit 7

PUTTING IT ALL TOGETHER
FOOD INVENTORY EVALUATION

You will want to present a well-prepared food diary analysis for this unit—in part because of what you will learn about your own eating habits! As you prepare your analysis, use the 11 topics listed below as guidelines. They will form the basis of your teacher's evaluation. You may also use the guidelines to evaluate your own performance.

Item	Points Possible	Self-Evaluation	Instructor Evaluation
Detailed Diary	20		
Research of Food Content	10		
Research of RDA	10		
Research of Vitamins	10		
Research of Minerals	10		
Research of Human Energy	10		
Summary of Diet and Average Daily Computation	5		
Graph	10		
Analysis and Conclusion	5		
Summary	5		
Bibliography (format)	5		
Total	**100**		

Appendix: Answers to Supplements

UNIT 1

B.6 Supplement—Ionic Formulas

	Cation	Anion	Formula	Name
1.	Ca^{2+}	O^{2-}	CaO	Calcium oxide
2.	Na^+	Cl^-	NaCl	Sodium chloride
3.	NH_4^+	NO_3^-	NH_4NO_3	Ammonium nitrate
4.	Cu^{2+}	OH^-	$Cu(OH)_2$	Copper(II) hydroxide
5.	Fe^{3+}	SO_4^{2-}	$Fe_2(SO_4)_3$	Iron(III) sulfate
6.	K^+	SO_3^{2-}	K_2SO_3	Potassium sulfite
7.	Na^+	PO_4^{3-}	Na_3PO_4	Sodium phosphate
8.	Pb^{2+}	Br^-	$PbBr_2$	Lead(II) bromide
9.	Pb^{2+}	CO_3^{2-}	$PbCO_3$	Lead(II) carbonate
10.	Al^{3+}	PO_4^{3-}	$AlPO_4$	Aluminum phosphate
11.	Mg^{2+}	HCO_3^-	$Mg(HCO_3)_2$	Magnesium hydrogen carbonate
12.	K^+	S^{2-}	K_2S	Potassium sulfide
13.	Ba^{2+}	SO_4^{2-}	$BaSO_4$	Barium sulfate
14.	Zn^{2+}	PO_4^{3-}	$Zn_3(PO_4)_2$	Zinc phosphate
15.	Fe^{3+}	Cl^-	$FeCl_3$	Iron(III) chloride

C.1 Supplement—Solubility Curves

1. 166 g
2. 42 g
3. 39 g
4. 84 °C
5. 30 °C
6. 23 °C
7. 20 g per 100 g H_2O
8. 11 g
9. 20 g
10. 134 g per 100 g H_2O
11. 18 g per 100 g H_2O
12. 133 g
13. 21 g
14. unsaturated
15. saturated
16. 53 °C
17. 8.0 mg
18. 3 mg per 1 Kg H_2O, or about 143%
19. 15 °C
20. 14 mg

C.2 Supplement—Solution Concentration

1. $(3.0 \text{ g}/100.0 \text{ g}) \times 100 = 3\%$
2. $(10.0 \text{ g}/50.0 \text{ g}) \times 100 = 20\%$
3. $(4.0 \text{ g}/39 \text{ g}) \times 100 = 10.3\%$
4. $(160 \text{ g}/260 \text{ g}) \times 100 = 62\% (61.5\%)$
5. $(0.008 \text{ g}/1000.008 \text{ g}) \times 100 = 0.0008\%$ or 8 ppm
6. $(0.0045 \text{ g}/1000.0045) \times 100 = 0.00045\%$, or 4.5 ppm
7. $550 \text{ g} \times 0.37 = 204 \text{ g}$
8. $500.0 \text{ g} \times (0.44/1\,000\,000) = 0.00022 \text{ g}$

UNIT 2

A.6 Supplement—Periodic Table and Predicting Formulas

Name	Symbol	Atomic Number	Atomic Mass	Number of Protons	Number of Neutrons	Number of Electrons
Boron	B	5	11	5	6	5
Zinc	Zn	30	65	30	35	30
Potassium	K	19	39	19	20	19
Titanium	Ti	22	48	22	26	22
Antimony	Sb	51	122	51	71	51
Uranium	U	92	238	92	146	92
Silver	Ag	47	108	47	61	47
Fermium	Fm	100	257	100	157	100
Platinum	Pt	78	195	78	117	78
Krypton	Kr	36	84	36	48	36
Radon	Rn	86	222	86	136	86

1. SiO_2
2. BaS
3. K_2S
4. BF_3
5. $LiBr$
6. SrO
7. InI_3
8. CaF_2
9. Al_2S_3
10. H_2O

B.3 SUPPLEMENT: RATES OF REACTION

1. The concentration of the five test tubes is:

Volume of Vinegar (mL)	Volume of Water Added (mL)	Molarity (M)
20	0	0.8
15	5	0.6
10	10	0.4
5	15	0.2
0	20	0.0

2. Higher concentrations of reactants react much more quickly than lower concentrations. The collision theory would suggest that when considering higher concentrations of reactant (which have more reacting particles in a given volume than lower concentrations), there are more effective collisions. This lab had a range of acid solutions from 0.8M to 0.0M. It would be expected, then, that the reaction for a 0.8M solution would be faster than the 0.2M solution. In this lab, when the chalk was placed in the vinegar, the chalk was observed to react, dissolve, and produce carbon dioxide gas, with a reaction duration of about 90 seconds. The chalk, when placed in the 0.2M solution, was observed to not completely react, still producing a steady, lower amount of carbon dioxide gas, with the reaction taking over 30 minutes.

3. Greater surface area of a reactant increases the rate of reaction. The collision theory would suggest that a reactant that has greater surface area than another would have a greater number of atoms or molecules open for reaction, thus increasing the rate of reaction. Crushing the tablet increased the reacting surface area, thus increasing the rate of reaction dramatically.

4. Higher temperatures increase the rate of reaction. Molecular-kinetics theory would suggest that increasing the temperature of the substances increases the rate of motion of the particles, resulting in more collisions.

5. The reactions in fireworks are very quick due to the high concentrations of explosive substances used, the great surface area of these powders, and the rapid release of heat promoting the reaction. A nail rusts slowly due to the low concentration of oxygen in the air (21%), the low surface area of the solid metal, and the relatively low temperatures of the environment.

6.

	Variables		
	Independent	**Dependent**	**Controlled (Some)**
Part 1: Concentration	Concentration of one reactant	Reaction completion time	• Volume • Temperature • Amount of second reactant • Solvent
Part 2: Surface area	Surface area of one reactant	Reaction completion time	• Volume • Temperature • Concentration • Amount of both reactants • Solvent
Part 3: Temperature	Temperature	Reaction completion time	• Volume • Concentrations • Reactant amounts

B.3 SUPPLEMENT 2: RATES OF REACTION

Teacher Notes

Student development of laboratory procedures can be very innovative and can lead them toward inquiry, meeting the intent of the National Science Education Standards. Alka Seltzer tablets and water are fairly safe, which enables the teacher to give students leeway to try different techniques to investigate how chemical reactions might be affected by concentration, temperature, and surface area. Here are some ideas that students might use to develop procedures for their investigations—although you should not limit the possibilities to these.

Analyzing concentration might be done by comparing how one tablet dissolves with dissolving two or more tablets. Investigating temperature might be done by evaluating the reaction in various temperatures of water (e.g., water at room temperature and water that is below room temperature). Investigating surface area might be accomplished by comparing how a whole tablet dissolves and one that has been powdered. Although these ideas are geared toward manipulating one of the reactants, students might realize that there are two reactants and that manipulating the conditions of either might provide evidence toward the inquiry being investigated.

Answers

1. Higher temperature increases the rate of reaction. Molecular-kinetics theory would suggest that increasing the temperature of the substances increases the rate of motion of the particles or molecules, resulting in more effective collisions.

2. Greater surface area of a reactant increases the rate of reaction. The reaction collision theory would suggest a reactant having greater surface area than another would have a greater number of atoms or molecules open for reaction, thus increasing the rate of reaction. Crushing the tablet increased the reacting surface area by thousands, thus increasing the rate of reaction dramatically.

3. The reactions in gaseous gasoline are very quick due to high concentrations of oxygen to gasoline molecules, the greater surface area of the gas, and the rapid release of heat promoting the reaction. Liquid gasoline does not burn (do not try this) due to the low (or no) concentration of oxygen in the gasoline, the low surface area of the liquid, and the relatively low temperatures of the environment.

C.1 Supplement—Atomic Inventory

1. A chemical equation is balanced if there are _____an equal number_____ of each kind of
_____element_____ on both sides of the equation.

2a. $CaCO_3 \rightarrow Ca$ __1__; C __1__; O __3__

   b. $(NH_4)_2SO_4 \rightarrow N$ __2__; H __8__; S __1__; O __4__

   c. $3H_2 \rightarrow H$ __6__

   d. $4Mg(OH)_2 \rightarrow Mg$ __4__; O __8__; H __8__

   e. $Ba(NO_3)_2 \rightarrow Ba$ __1__; N __2__; O __6__

3a. $2 Na + 2 H_2O \rightarrow 2 NaOH + H_2$

   __2__ Na __2__

   __4__ H __4__

   __2__ O __2__   Balanced? *Yes*

   b. $4 NH_3 + 6 NO \rightarrow 5 N_2 + 6 H_2O$

   __10__ N __10__

   __12__ H __12__

   __6__ O __6__   Balanced? *Yes*

4a. __1__ Na __1__

   __1__ Cl __2__

   __2__ F __1__   Balanced? *No*

   b. __3__ Na __3__

   __3__ Br __2__

   __3__ H __2__

   __1__ P __1__

   __4__ O __4__   Balanced? *No*

c. _4_ N _6_

    _4_ H _8_

    _4_ O _4_   Balanced? *No*

d. _4_ Ag _4_

    _8_ H _8_

    _4_ S _2_

    _2_ O _4_   Balanced? *No*

e. _2_ Bi _2_

    _6_ F _6_   Balanced? *Yes*

f. _1_ Al _1_

    _1_ Ni _1_

    _2_ N _3_

    _6_ O _9_   Balanced? *No*

g. _3_ Na _3_

    _7_ B _7_

    _12_ H _12_

    _12_ F _12_   Balanced? *Yes*

h. _12_ C _12_

    _20_ H _20_

    _12_ N _12_

    _36_ O _36_   Balanced? *Yes*

i.

	10	Ca	10
	2	F	2
	6	P	6
	52	O	52
	14	H	14
	7	S	7

Balanced? *Yes*

C.2 Supplement—Balancing Equations

1. $2\,H_2 + O_2 \rightarrow 2\,H_2O$
2. $2\,Mg + O_2 \rightarrow 2\,MgO$
3. $Ca + 2\,H_2O \rightarrow Ca(OH)_2 + H_2$
4. $Cu + 2\,HgNO_3 \rightarrow Cu(NO_3)_2 + 2\,Hg$
5. $C_3H_8 + 5\,O_2 \rightarrow 3\,CO_2 + 4\,H_2O$
6. $2\,Al + 3\,F_2 \rightarrow 2\,AlF_3$
7. $4\,Fe + 3\,O_2 \rightarrow 2\,Fe_2O_3$
8. $Fe_3O_4 + 4\,H_2 \rightarrow 3\,Fe + 4\,H_2O$
9. $4\,HBr + O_2 \rightarrow 2\,Br_2 + 2\,H_2O$
10. $Al_2O_3 + 6\,HCl \rightarrow 2\,AlCl_3 + 3\,H_2O$
11. $3\,NH_4OH + FeCl_3 \rightarrow Fe(OH)_3 + 3\,NH_4Cl$
12. $4\,NH_3 + 5\,O_2 \rightarrow 4\,NO + 6\,H_2O$
13. $I_2 + 10\,HNO_3 \rightarrow 2\,HIO_3 + 10\,NO_2 + 4\,H_2O$
14. $3\,CaO + P_2O_5 \rightarrow Ca_3(PO_4)_2$
15. $6\,NaOH + Al_2(SO_3)_3 \rightarrow 3\,Na_2SO_3 + 2\,Al(OH)_3$

C.3 Supplement—Molar Mass

1. 28 g/mol
2. 58 g/mol
3. 342 g/mol
4. 184 g/mol
5. 222 g/mol
6. 346 g/mol
7. 6.3 mol $\times$ 28 g/mol = 180 g (176 g)
8. 84.6 g / 58 g/mol = 1.5 mol (1.46 mol)
9. 564 g / 342 g/mol = 1.65 mol
10. 3.95 mol $\times$ 184 g/mol = 727 g (726.8 g)
11. 0.985 g / 222 g/mol = 0.00444 mol
12. 36.5 mol $\times$ 346 g/mol = 12 600 g (12 629 g)

C.4 Supplement—Percent Composition and Conservation of Mass

1. 13%
2. 40%
3. 51%
4. 64%
5. Al_2O_3 is 53% aluminum, while $Al(NO_3)_3$ is 12%

Balance these equations:

6. $2\,Al \ + \ 6\,HCl \rightarrow \ 2\,AlCl_3 \ + \ 3\,H_2$

$2(27) \ + \ 6(36) \ = \ 2(132) \ + \ 3(2)$

$54 \quad + \ 216 \quad = \ 264 \quad + \ 6$

$\qquad\qquad 270 \ = \ 270$

7. $3\,F_2 \ + \ 2\,NiI_3 \rightarrow 2\,NiF_3 \ + \ 3\,I_2$

$3(38) \ + \ 2(440) = \ 2(116) \ + \ 3(254)$

$114 \ + \ 880 \quad = \ 232 \quad + \ 762$

$\qquad\qquad 994 \quad = \ 994$

8. $4\,Fe \ + \ 3\,O_2 \ \rightarrow \ 2\,Fe_2O_3$

$4(56) \ + \ 3(32) \ = \ 2(160)$

$224 \ + \ 96 \quad = \ 320$

$\qquad\quad 320 \quad = \ 320$

9. $3\,NH_4OH + \ FeCl_3 \ \rightarrow \ 3\,NH_4Cl \ + \ Fe(OH)_3$

$3(35) \qquad + \ 1(161) = \ 3(53) \qquad + \ 1(107)$

$105 \qquad + \ 161 \quad = \ 159 \quad + \ 107$

$\qquad\qquad 266 \quad = \ 266$

UNIT 3

A.5 Supplement—Electron Configuration

Name	Symbol	Atomic Number	Number of Electrons in Outermost Energy	Electron Dot Notation
1. Hydrogen	H	1	1	H·
2. Carbon	C	6	4	·Ċ·
3. Oxygen	O	8	6	·Ö:
4. Nitrogen	N	7	5	·Ṅ:
5. Chlorine	Cl	17	7	:Ċl:
6. Calcium	Ca	20	2	Ca:
7. Iron	Fe	26	2	Fe:
8. Copper	Cu	29	1	Cu·
9. Selenium	Se	34	6	·Ṡe:
10. Krypton	Kr	36	8	:Ḳr:
11. Xenon	Xe	54	8	:Ẍe:
12. Gold	Au	79	1	Au·
13. Astatine	At	85	7	:Ȧt:
14. Radon	Rn	86	8	:Ṙn:
15. Silver	Ag	47	1	Ag·

A.7 Supplement—Naming Branched Alkanes

1. 2-methyloctane
2. 3-ethylheptane
3. 2-methylbutane
4. 4,6-dimethylnonane
5. $CH_3-CH-CH_2-CH_2-CH_3$
    $\quad\quad\; | $
    $\quad\quad CH_3$

6. $CH_3-CH_2-CH-CH_2-CH_2-CH_2-CH_2-CH_2-CH_2-CH_3$
    $\quad\quad\quad\quad\;\; |$
    $\quad\quad\quad\quad CH_2-CH_3$

7. $CH_3-CH_2-CH_2-CH-CH_2-CH_2-CH_3$
$|$
$CH_2-CH_2-CH_3$

8. $CH_3-CH-CH-CH_2-CH_2-CH_2-CH_2-CH_3$
$||$
$CH_3\ CH_3$

B.4 Supplement—Heats of Combustion

1a. $\dfrac{2660\ g}{1\ gal} \times \dfrac{47.8\ kJ}{1\ g} = \dfrac{127\ 000\ kJ}{gal}$

b. $\dfrac{127\ 000\ kJ}{gal} \times \dfrac{1\ gal}{22.0\ miles} = \dfrac{5770\ kJ}{mile}$

c. $5770\ kJ/mile \times 75\% = 4330\ kJ/mile$

2a. $2\ 800\ 000\ cal \times \dfrac{1\ g}{4000\ cal} = 700\ g$

b. $2\ 800\ 000\ cal \times \dfrac{1\ g}{9000\ cal} = 311\ g$

c. $12\ 000\ kJ \times \dfrac{1\ g}{47.8\ kJ} = 251\ g \qquad 251\ g \times \dfrac{1\ gal}{2660\ g} = 0.0944\ gal$

3a. $16\ 500\ pounds \times \dfrac{1\ acre}{1900\ pounds} = 8.68\ acres$

b. $16\ 500\ pounds \times \dfrac{1\ person}{400\ pounds} = 41\ persons$

c. $1\ pound \times \dfrac{1\ year}{16\ 500\ pounds} \times \dfrac{15\ 000\ miles}{1\ year} = 0.91\ miles$

d. $135\ 000\ 000\ cars \times \dfrac{8.68\ acres}{1\ car} = 1\ 170\ 000\ 000$ acres, or $1\ 830\ 000$ sq mi, or $4\ 739\ 678$ sq km;
this is more than 3x the area of Alaska, or over 30x the area of Michigan

C.4 Supplement—The Builders

1. alcohol

2. ester

3. carboxylic acid

4. alcohol

5. ether

6. ester

7. ether

8. alcohol

9. carboxylic acid

10. none of the above (an alkane)

A.4 SUPPLEMENT: COMPRESSIBILITY OF MATERIALS

1. Students should be aware from previous learning about the incompressibility of liquids and solids, and the compressibility of gases. Therefore they should define the air as compressible, salt as incompressible, water as incompressible, and the second liquid as incompressible. Given that basis, increasing the pressure on any of the samples should only cause a significant change in volume of the air, which, if it remained a gas under 1000 atmospheres, would have about 1/1000th of its original volume.

2. Gravity affects the shape of the liquid in the syringe, providing a force that causes it to flow throughout the container and flattens out the surface. If there is a slight curvature to the surface of the liquid, there may be some adhesion to the sides of the syringe coupled with intermolecular forces.

3. The solid cylinder of copper would be expected to retain its own form (not flow) and remain incompressible.

4. The air (gas) and two liquids are fluids. Although the salt crystals may be poured into the syringe, they do not flow (such as the others if the tip were opened), and as such, salt is not a fluid.

5. Students should make the transition from their observations that, generally, liquids and solids are relatively incompressible, and that gases are compressible.

6. Student drawings should show that the volumes of the liquids and solid did not change and that the molecules remained spaced about the same distance apart. The air would be depicted as a reduced volume after compression, with the molecules much closer together.

UNIT 4

A.5 Supplement—Boyle's Law

1. 106 mL (10.50 kPa × 100.0 mL = 9.91 kPa × V)

2. 992 mm Hg (731 mm Hg × 95 cm^3 = P × 70.0 cm^3)

3a. 25 m^3 (P × 50.0 m^3 = 2P × V)

b. 100 m^3 (P × 50.0 m^3 = .5P × V)

c. 16.7 m^3 (P × 50.0 m^3 = 3P × V)

4. The balloon will increase in volume to 1 ft^3 (4 atm × .25 ft^3 = 1 atm × V)

5. 1.5 atm (V × 3 atm = 2V × P)

6. 30.6 psi (125 kPa × 760 cm^3 = P × 450 cm^3) $\left(211.1 \text{ kPa} \times \dfrac{14.7 \text{ psi}}{101.3 \text{ kPa}} = 30.6 \text{ psi}\right)$

A.7 Supplement—Charles' Law

1. 546 K or 273 °C (V/273 K = 2V/T)

2a. 20 m^3 (10.0 m^3/273 K = V/546 K)

b. 2.5 m^3 (10.0 m^3/273 K = V/68.3 K)

3. 542 mL (500 mL/298 K = V/323 K)

4. 281 K or 8 °C (560 cm^3/393 K = 400 cm^3/T)

5. The gas will increase 27 mL to 127 mL (100.0 mL/263 K = V/333 K)

6. 262 K or −11 °C (2.75 L/293 K = 2.46 L/T)

A.8 Supplement—Temperature-Pressure

1. 240 kPa (199 kPa/293 K = P/353 K)

2. 1810 psi (2000 psi/324 K = P/293 K)

3. 527 K or 254 °C (0.700 atm/295 K = 1.25 atm/T)

4. 397 K or 124 °C (3.00 atm/298 K = 4.00 atm/T)

5. 27.0 in Hg (29.92 in Hg/303 K = P/273 K); if the can had not collapsed, the pressure difference on the cap would have made it difficult to open the can.

6. .12 atm change (86.0 °F = 30 °C; 1.75 atm/303 K = P/283 K; change = 1.75 − P = 1.75 − 1.63)

A.9 Supplement—Molar Volume

1. 56.0 L (Volume = 2.50 mol × 22.4 L/1 mol)

2. 3.01 moles (# moles SO_2 = 67.4 L × 1 mol/22.4 L)

3a. $H_2 + Cl_2 \rightarrow 2HCl$

b. 0.875 L (Volume = 1.75 L HCl × 1 mol H_2/2 mol HCl × 22.4 L H_2/1 mol H_2 × 1 mol/22.4 L HCl)

c. 8.65 moles (# moles = 8.65 mol H_2 × 1 mol H_2/1 mol Cl_2)

4a. $C_3H_8 + 5 O_2 \rightarrow 3 CO_2 + 4 H_2O$

b. 1.75 L (0.350 L C_3H_8 × 5 mol O_2/1 mol C_3H_8 × 22.4 L O_2/1 mol O_2 × 1 mol C_3H_8/22.4 L)

c. 1.40 L H_2O (0.350 L C_3H_8 × 4 mol H_2O/1 mol C_3H_8

5. 0.028 moles neon gas (radius = 1.30 cm; Volume = $\pi(1.3 \text{ cm})^2$ × 120 cm = 637 cm^3 = 0.637 L; # of moles = 0.637 L Ne gas × 1 mole Ne/22.4 L)

UNIT 5

C.1 Supplement—Electrochemical Changes

1. $Cu + 4 HNO_3 \rightarrow Cu(NO_3)_2 + 2 NO_2 + 2 H_2O$
   Copper was oxidized (oxidation number changed from 0 to +2);
   nitrogen was reduced (oxidation number changed from +5 to +4).

2a. $Fe + 2 HCl \rightarrow FeCl_2 + H_2$

 b. Yes

 c. Iron was oxidized ($0 \rightarrow +2$), and hydrogen was reduced ($+1 \rightarrow 0$).

3a. $NaOH + HCl \rightarrow NaCl + H_2O$

 b. No

 c. The oxidation number for all of the elements of the reaction did not change.

   (Na = +1, O = −2, H = +1, Cl = −1)

4. The copper(II) ions are reduced by the magnesium, *which is oxidized.* The magnesium is less electronegative so the copper(II) ions have a greater attraction for the shared electrons. The overal reaction is:
   $Cu^{+2} + Mg \rightarrow Cu + Mg^{+2}$

C.3 Supplement—Building Skills: Getting a Charge from Electrochemistry

1. $2(Cr \rightarrow Cr^{+3} + 3 e^-)$
   $3(Pb^{+2} + 2 e^- \rightarrow Pb)$

   $2 Cr \rightarrow 2 Cr^{+3} + 6 e^-$
   $3 Pb^{+2} + 6 e^- \rightarrow 3 Pb$
   _____
   $\mathbf{3\,Pb^{+2} + 2\,Cr \rightarrow 3\,Pb + 2\,Cr^{+3}}$

2. $Ag^+ + 1e^- \rightarrow Ag$
   $Fe \rightarrow Fe^{+3} + 3 e^-$

   $3 Ag^+ + 3 e^- \rightarrow 3 Ag$
   $Fe \rightarrow Fe^{+3} + 3 e^-$
   _____
   $\mathbf{3Ag^+ + Fe \rightarrow 3Ag + Fe^{+3}}$

3. $K \rightarrow K^+ + 1 e^-$
   $2 H_2O + 2 e \rightarrow H_2 + 2 OH$

   $2 K \rightarrow 2 K^+ + 2 e^-$
   $2 H_2O + 2 e^- \rightarrow H_2 + 2 OH$
   _____
   $\mathbf{2\,H^+ + 2\,K \rightarrow 2\,H + 2\,K^+}$ (or $H^2 + 2\,K^+$)

4. $MnO^{4-} + 8 H^+ + 5 e^- \rightarrow$
   $Mn^{2+} + 4 H_2O$

   $5 Sn^{2+} \rightarrow 5 Sn^{4+} + 10 e^-$
   $2 MnO^{4-} + 16 H^+ + 10 e^- \rightarrow 2 Mn^{2+} + 8 H_2O$
   _____
   $\mathbf{2\,MnO^{4-} + 5\,Sn^{2+} + 16\,H^+ \rightarrow 2\,Mn^{2+} + 5\,Sn^{4+} + 8\,H_2O}$

5. $Cd^{+2} + 2 e^- \rightarrow Cd$
   $Ni \rightarrow Ni^{+2} + 2 e^-$

UNIT 6

A.4 Supplement—Building Skills

1a. Most of the particles passed through without being deflected.

b. Some of the positive alpha particles were deflected.

c. Most of the particles passed through without being deflected.

2. Isotopes of the same element have the same number of protons and electrons (if they have the same charge). They also have the same atomic number. They are different by the different number of neutrons and different atomic mass.

3. $^{52}_{24}$ Cr, Chromium-52; 24 protons, 28 neutrons, 24 electrons

4. $^{201}_{80}$ Hg

5a. $^{59}_{27}$ Co^{2+}

b. $^{127}_{53}$ I$^-$

6a. F; **b.** T; **c.** F; **d.** T; **e.** F; **f.** T

7a. 24; **b.** 148; **c.** 157; **d.** 112

A.5 Supplement—Building Skills: Finding Average Atomic Weight

1. 10.802

2. 63.611

3. 69.7946

4. 79.9916

5. 85.558

6. 35.454

7. 107.9656

8. 28.11

9. 24.325

10. 118.839

B.5 Supplement—Building Skills

1. $^{87}_{36}$ Kr $\rightarrow$ $^{87}_{37}$ Rb $+$ $^{0}_{-1}$ e

2. $^{240}_{96}$ Cm $\rightarrow$ $^{236}_{94}$ Pu $+$ $^{4}_{2}$ He

3. $^{232}_{92}$ U $\rightarrow$ $^{228}_{90}$ Th $+$ $^{4}_{2}$ He

4. $^{32}_{14}$ Si $\rightarrow$ $^{32}_{15}$ P $+$ $^{0}_{-1}$ e

5. $^{71}_{30}$ Zn $\rightarrow$ $^{71}_{31}$ Ga $+$ $^{0}_{-1}$ e

6. $^{243}_{95}$ Am $\rightarrow$ $^{239}_{93}$ Np $+$ $^{4}_{2}$ He

7. $^{60}_{27}$ Co $\rightarrow$ $^{60}_{28}$ Ni $+$ $^{0}_{-1}$ e

8. $^{32}_{15}$ P $\rightarrow$ $^{32}_{16}$ S $+$ $^{0}_{-1}$ e

9. $^{150}_{64}$ Gd $\rightarrow$ $^{146}_{62}$ Sm $+$ $^{4}_{2}$ He

10. $^{210}_{82}$ Pb $\rightarrow$ $^{206}_{81}$ Tl $+$ $^{0}_{-1}$ e $+$ $^{4}_{2}$ He

C.1 Supplement—Building Skills: Half-Lives

1a. 12.5%

  b. 5 half-lives

2.

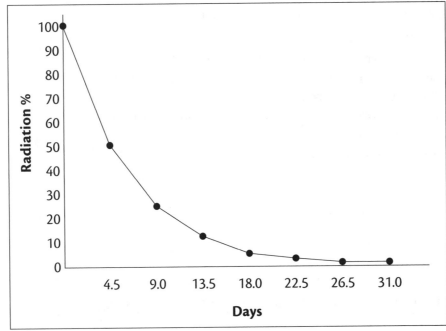

2a. approximately 11%

  b. approximately 15 days

  c. 99.2%

3. about 8 hours

4. 115.2 years

5. 0.117 counts

6. 22 920 years old

7. 400 counts

C.4 Supplement—Building Skills: Artificial Nuclear Reaction Equations

1. $^{10}_{5}B + ^{1}_{0}n \rightarrow ^{7}_{3}Li + ^{4}_{2}He$

2. $^{9}_{4}Be + ^{1}_{1}p \rightarrow ^{6}_{3}Li + ^{4}_{2}He$

3. $^{253}_{99}Es + ^{4}_{2}He \rightarrow ^{256}_{101}Md + ^{1}_{0}n$

4. $^{7}_{3}Li + ^{1}_{1}p \rightarrow ^{7}_{4}Be + ^{1}_{0}n$

5. $^{241}_{94}Pu + ^{1}_{0}n \rightarrow ^{242}_{94}Pu + \gamma$

6. $^{40}_{18}Ar + ^{4}_{2}He \rightarrow ^{43}_{20}Ca + ^{1}_{0}n$

7. $^{252}_{99}Es + ^{9}_{4}Be \rightarrow ^{258}_{103}Lr + 3\,^{1}_{0}n$

8. $^{238}_{92}U + ^{4}_{2}He \rightarrow ^{239}_{94}Pu + 3\,^{1}_{0}n$

9. $^{235}_{92}U + ^{1}_{0}n \rightarrow ^{137}_{52}Te + ^{97}_{40}Zr + 2\,^{1}_{0}n$

10. $^{3}_{2}He + ^{3}_{2}He \rightarrow ^{4}_{2}He + 2\,^{1}_{1}H$

UNIT 7

PIAT Supplement: What Kind of Food Am I?

1a. $67 \text{ g} - 8 \text{ g} = 59 \text{ g}$

b. $45 \text{ g} + 8 \text{ g} = 53 \text{ g}$; $(53 \text{ g}/67 \text{ g}) \times 100 = 79\%$

2a. Chicken Delight has a smaller percent of the daily value for fat, saturated fat, cholesterol, and sodium. It also has a high percentage of vitamin A and iron.

b. Chicken Supreme has lower Calories and higher percents of vitamin C and calcium.

3. $15 \text{ mg} \times 15\% = 2.25 \text{ mg}$

4. $(23 \text{ g}/44 \text{ g}) \times 100 = 52.3\%$; $(23 \text{ g}/59 \text{ g}) \times 100 = 39\%$

5. $(70/370) \times 100 = 18.9\%$

6. $(130/290) \times 100 = 44.8\%$

7. Energy is primarily obtained from carbohydrates and fats. Vitamins are present in only small amounts.

8. Everything in food is "chemical." When people say no chemicals, they often mean no additives, but this is incorrect.

PIAT Supplement: Using Food Label Information

1. $(25/120) \times 100 = 21\%$

2a. Unsaturated

b. Soybean oil

3. $23 \text{ g} \times 4 \frac{\text{Cal}}{\text{g}} = 92 \text{ Cal}$

4. $2 \text{ g} \times 4 \frac{\text{Cal}}{\text{g}} = 8 \text{ Cal}$; $\frac{8 \text{ Cal}}{120 \text{ Cal}} \times 100 = 6.7\%$

5. $\frac{60 \text{ mg}}{\text{Day}} \times 0.25 = 15 \text{ mg}$ (Use RDA, Student book, page 525)

6. $1200 \times .06 = 72 \text{ mg}$ (Use RDA, page 545)

7. Most phosphorous is found in animal protein.

8a. folic acid, niacin, riboflavin, vitamin C, and B_6, thiamin

b. vitamin D, A, and E (in ingredients)

9a. calcium, magnesium, phosphorous, sodium, and potassium

b. zinc, copper, and iron

10. whole oat flour

11. BHT; antioxidant

12. sodium sulfite; Na_2SO_3

13. rounding of figures